Chemical Sensors for Hostile Environments

Volume 130

Chemical Sensors for Hostile Environments

Proceedings of the Chemical Sensors for Hostile Environments symposium, held at the 103rd Annual Meeting of The American Ceramic Society April 22–25, 2001, in Indianapolis, Indiana.

Edited by

G.M. Kale
University of Leeds

S.A. Akbar
The Ohio State University

M. Liu
Georgia Institute of Technology

Published by
The American Ceramic Society
735 Ceramic Place
Westerville, Ohio 43081
www.ceramics.org

Proceedings of the Chemical Sensors for Hostile Environments symposium, held at the 103rd Annual Meeting of The American Ceramic Society April 22–25, 2001, in Indianapolis, Indiana.

Copyright 2002, The American Ceramic Society. All rights reserved.

Statements of fact and opinion are the responsibility of the authors alone and do not imply an opinion on the part of the officers, staff, or members of The American Ceramic Society. The American Ceramic Society assumes no responsibility for the statements and opinions advanced by the contributors to its publications or by the speakers at its programs. Registered names and trademarks, etc., used in this publication, even without specific indication thereof, are not to be considered unprotected by the law.

No part of this book may be reproduced, stored in a retrieval system, or transmitted in any form or by any means, electronic, mechanical, photocopying, microfilming, recording, or otherwise, without written permission from the publisher.

Authorization to photocopy for internal or personal use beyond the limits of Sections 107 and 108 of the U.S. Copyright Law is granted by the American Ceramic Society, ISSN 1040-1122 provided that the appropriate fee is paid directly to the Copyright Clearance Center, Inc., 222 Rosewood Drive, Danvers, MA 01923 USA, www.copyright.com. Prior to photocopying items for educational classroom use, please contact Copyright Clearance Center, Inc.

This consent does not extend to copying items for general distribution or for advertising or promotional purposes or to republishing items in whole or in part in any work in any format.

Please direct republication or special copying permission requests to the Senior Director, Publications, The American Ceramic Society, PO Box 6136, Westerville, Ohio 43086-6136, USA.

Cover photo: "Recovered GaO(OH) precipitates without urea" is courtesy of A. Cüneyt Taş, Peter J. Majewski, and Fritz Aldinger, and appears as figure 5(left) in their paper "Synthesis of Gallium Oxide Hydroxide Crystals in Aqueous Solutions with or without Urea and their Calcination Behavior," which begins on page 105.

For information on ordering titles published by The American Ceramic Society, or to request a publications catalog, please call 614-794-5890, or visit www.ceramics.org.

4 3 2 1–05 04 03 02

ISSN 1042-1122
ISBN 1-57498-138-2

Contents

Solid-State Electrochemical Sensors for Automotive Applications
R. Mukundan, E.L. Brosha, and F.H. Garzon 1

Zirconia-Based Potentiometric NO_x Sensor Utilizing Pt and Au Electrodes
D.J. Kubinski, J.H. Visser, R.E. Soltis, M.H. Parsons, K.E. Nietering, and S.G. Ejakov 11

Packaging Planar Exhaust Sensors for Hostile Exhaust Environments
C.S. Nelson 19

Importance of Gas Diffusion in Semiconductor Gas Sensors
N. Yamazoe, G. Sakai, and N. Matsunaga 27

Durability of Thick-Film Ceramic Gas Sensors
A. Merhaba, S. Akbar, B. Feng, G. Newaz, L. Riester, and P. Blau 37

Preparation and Characterization of Indium-Doped Calcium Zirconate for the Electrolyte in Hydrogen Sensors for Use in Molten Aluminum
A.H. Setiawan and J.W. Fergus 47

Antimony Sensors for Molten Lead Using K-β-Al_2O_3 Solid Electrolytes
R. Kurchania and G.M. Kale 57

Preparation and Characterization of Iron Oxide–Zirconia Nanopowder for Its Use as an Ethanol Sensor Material
C.V.G. Reddy, S.A. Akbar, W. Cao, O.K. Tan, and W. Zhu 67

Synthesis and Characterization of 2–3 Spinels as Material for Methane Sensors
S. Poomiapiradee, R.M.D. Brydson, and G.M. Kale 79

Ammonia and Alcohol Gas Sensors Using Tungsten Oxide
M. Sriyudthsak, S. Udomratananon, and S. Supothina 91

Low-Temperature Gas Sensing Using Laser Activation
M. Sriyudthsak and V. Rungsaiwatana 97

Synthesis of Gallium Oxide Hydroxide Crystals in Aqueous Solutions with or without Urea and Their Calcination Behavior
A.C. Taş, P.J. Majewski, and F. Aldinger 105

Index 115

Preface

Chemical sensors have been recognized as one of the important components of the process control circuit in a chemical and metallurgical industry. Chemical sensors play a significant role in the field of environmental management, risk assessment, and health. The conception of an idea to the development of a functional device involves researchers from physics, chemistry, materials, electrical and electronics, software, and engineering, which makes the area of chemical sensors truly multidisciplinary and challenging. A symposium on chemical sensors for hostile environments was held in Indianapolis, Indiana, April 22–25, 2001. The symposium was part of the 103rd Annual Meeting of The American Ceramic Society and was co-sponsored by the Basic Science and Engineering Ceramic divisions and the National Institute of Ceramic Engineers.

The primary objectives of the symposium were to present novel materials, techniques, applications, and challenges of chemical sensors technology; emphasize the multidisciplinary nature of the chemical sensors research; and promote exchange of ideas and networking between researchers and end users of the technology. We hope that the symposium, the first in a series, maintains a sustained interest in this fertile and technologically demanding area of research. The current volume consists of selected papers from a total of 28 that were presented at this symposium by researchers from 10 different countries, making the symposium not only a multidisciplinary but also an international event as well.

We are extremely grateful to all the contributors for making the symposium a success. We would like to thank the authors for their efforts and the reviewers for their expertise and valuable time. We would also like to acknowledge the time and efforts of the session chairs and co-chairs. Special thanks are extended to all the invited speakers, professors Prabir Dutta, Noboru Yamazoe, Harry Tuller, Jeffrey Fergus, Sheikh Akbar, and Drs. Garry Hunter and Rangachary Mukundan, who enlightened the audience with some exciting developments and future trends in the area of chemical sensors technology.

G. M. Kale

S. A. Akbar

M. Liu

SOLID-STATE ELECTROCHEMICAL SENSORS FOR AUTOMOTIVE APPLICATIONS

R. Mukundan, E. L. Brosha and F. H. Garzon
Los Alamos National Laboratory
MS D429, SM-40, TA-3
MST –11
Los Alamos
NM 87545

ABSTRACT

Several types of electrochemical sensors have been studied for application in an automobile. The most prevalent amongst these is the zirconia oxygen sensor, which is also known as the λ–sensor or Exhaust Gas Oxygen (EGO) sensor. This sensor is used to monitor the oxygen partial pressure in the exhaust, which in turn is utilized to control the air/fuel ratio to the engine. Currently, two such oxygen sensors are being used to determine the "state-of-health" of the catalyst. One sensor is placed upstream of the catalyst and another downstream, and the catalyst efficiency is indirectly monitored by measuring its' oxygen storage capacity. However such a system has several disadvantages and works only when the engine is operated around stoichiometry. There is increasing interest in developing a hydrocarbon sensor that can be used directly to measure the amount of non-methane hydrocarbons (NMHC) in the exhaust stream. The results of a newly developed mixed-potential sensor using platinum (Pt) and $La_{0.8}Sr_{0.2}CrO_3$-oxide (LSCO) electrodes on a 8m% yttria stabilized zirconia (YSZ) electrolyte are presented and its potential application as an automotive NMHC sensor is discussed. This sensor has a response of 40 mV to 500 ppm of propylene at 773K in $1\%O_2$ while the response to the same amount of CO is only 2 mV.

INTRODUCTION

Tailpipe emissions from automobiles are under increasing scrutiny of the Environmental Protection Agency (EPA) and California Air Resource Board (CARB). California exhaust emission standards for all 2001 model year passenger cars mandates that the non methane organic gasses (NMOG) coming out of the tailpipe of Tier 1 and LEV vehicles be less than 0.25 and 0.075 g/mile (1.55 x 10^{-4} and 4.66 x 10^{-5} g/m) respectively. Moreover the EPA and CARB also

---

To the extent authorized under the laws of the United States of America, all copyright interests in this publication are the property of The American Ceramic Society. Any duplication, reproduction, or republication of this publication or any part thereof, without the express written consent of The American Ceramic Society or fee paid to the Copyright Clearance Center, is prohibited.

require manufacturers to comply with On-Board Diagnostic (OBD-II) requirements, which mandate that the automobile user be warned by a Malfunction Indicator Light (MIL) when the tailpipe emissions increase by 0.4 g/mile (2.5 x 10^{-4} g/m). These requirements are currently being met by the dual EGO sensor technology.[1,2]

The dual EGO sensor consists of two **Heated-EGO** (HEGO) sensors, one placed up-stream of the catalyst and the other down-stream. The up-stream sensor monitors the oxygen partial pressure in the exhaust gases before the catalytic convertor and is used to adjust the Air/Fuel (A/F) ratio to the engine. If the engine is run around stoichiometry, then this sensor displays a step-like response when the A/F ratio changes from rich to lean and vice-versa.[3] However, the response of the second sensor (down-stream of the catalyst) is either delayed and/or dampened due to the oxygen storage capacity of the convertor. The precious metals and the ceria-washcoat in the catalyst have a tendency to store oxygen, which is consumed during hydrocarbon (HC) oxidation and replenished during nitrogen oxide (NOx) reduction. This oxygen storage capacity (OSC) can be linked to the HC conversion efficiency of the catalyst and is found to have a step like relationship where the OSC decreases dramatically when the HC conversion efficiency drops from around 95% to 80%. Therefore the ratio of the two sensor responses can be used as an indirect measure of the hydrocarbon conversion efficiency of the catalyst.[1,2]

Although this dual oxygen sensor approach has worked successfully, there are several disadvantages to this dual EGO sensor approach because it is not a direct measure of the HC conversion efficiency of the catalyst. The ratio of the responses of the two HEGO sensors is very sensitive to the HC conversion efficiency when the efficiency is > 80%. However, this measure is insensitive when the HC conversion efficiency drops below 70%.[1] Hence this method tends to classify catalysts that are working at 60-75% efficiency as "failed catalysts". Moreover, in order for this technology to work, the engine should be operated in the closed loop fuel control mode that results in the oscillations of the A/F ratio around stoichiometry. These problems can be greatly alleviated by the development of a sensor that would directly measure the HC content before and after the catalyst.

One such technology that has the potential to measure HCs in a background of O_2, N_2, H_2O, CO_2, CO and NOx is the mixed-potential sensor.[4,5] Mixed potential sensors have been used in the past for the detection of CO[6,7], NOx[8], H_2[9] and HCs.[10] These sensors rely on the fact that two dissimilar electrodes on an oxygen-ion conducting electrolyte exhibit different non-equilibrium potentials in the presence of a reducing gas and oxygen. The potential at any one electrode is fixed by the rate of electrochemical oxidation of the reducing gas (CO, HC's or NO_2) and the rate of electrochemical reduction of oxygen (Eqns. 1 & 2).[4]

$$\frac{1}{2}O_2 + V_o^{\cdot\cdot} + 2e^{'} \leftrightarrow O_o \quad (1)$$

$$CO + O_o \leftrightarrow CO_2 + V_o^{\cdot\cdot} + 2e^{'} \quad (2)$$

The factors that determine the mixed-potential at an electrode/electrolyte interface include; the electrode material, electrode surface-area, 3-phase interface length, amount of reducing gas and oxygen present, flow rates of the gases, and temperature of the measurement.[11] The working principle of a mixed potential sensor operating at 773-873 K in a 1%O_2 stream with 0-500 ppm of reducing gas was recently illustrated using a four electrode "Pt/$Ce_{0.8}Gd_{0.2}O_{1.9}$/Au" sensor consisting of Pt and Au sensing electrodes and 2 corresponding Pt reference electrodes.[12] It was found that the steady state current was determined primarily by the amount of electrochemical oxidation of the reducing-gas while the voltage was determined primarily by the oxygen reduction kinetics of the Au electrode. The response of this sensor to various gases is illustrated in Figure 1 while the stability of response of the individual Au and Pt electrodes is represented in Figure 2.[12] Figure 1 shows that the Au electrode has very little selectivity to HCs relative to CO and H_2. Moreover, from Figure 2 it is clear that the mixed potential is primarily at the Au electrode while the Pt electrode acts like a psedo-reference electrode and the stability of the Au electrode is less than desirable.

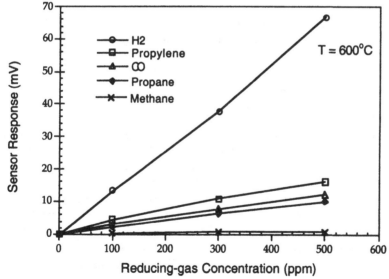

Figure 1. Response of a "Pt/$Ce_{0.8}Gd_{0.2}O_{1.9}$/Au" sensor to various reducing-gases at 873 K. Reproduced by permission of The Electrochemical Society, Inc.[12]

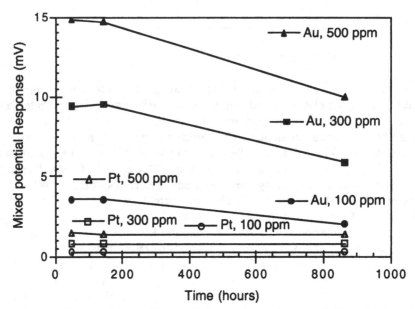

Figure 2. Stability of the gold and platinum electrode potentials to CO in a 1%O_2 stream at 873 K. Reproduced by permission of The Electrochemical Society, Inc.[12]

Recently, there have been several publications trying to address these limitations by replacing the Au electrode with an oxide or composite electrode. The electrode materials that have been reported include Au-10%Ta_2O_5[10], In_2O_3-0.1%MnO_2,[13] and CdO.[14] In this paper we report the response of a $La_{0.8}Sr_{0.2}CrO_3$ (LSCO) electrode on a YSZ electrolyte.

EXPERIMENTAL

The sensor consists of a LSCO electrode, a Pt electrode and a YSZ electrolyte. The LSCO electrode was prepared from $La_{0.8}Sr_{0.2}CrO_3$ powders from Praxair®. The powders were first uniaxially pressed in a 1.9 cm die at 15 Mpa for 300 s and then isostatically pressed at 200 MPa for 300 s. The LSCO samples were then sintered at 1923 K for 36000 s and pellets of approximately 2x2x2 mm^3 were cut from this ceramic to be used as the LSCO electrode. A LSCO pellet and a Pt wire (dia. = 0.25 mm) were then placed on a 1.9 cm die and covered with 3 gms of YSZ powder. This was then pressed at 23 Mpa for 300 s and the resulting sensor assembly was sintered at 1273 K for 36000 s.[15] Two identical sensors (sensor #1 and sensor #2) were prepared using the above technique.

The sensor was then mounted on an alumina rod with Pt leads which was then placed in a quartz tube and heated in a furnace to the operating temperature. The flow rates of the various gas mixtures were controlled using automatic MKS mass flow controllers while the voltage from the sensor was monitored using a Keithly 2400 Source Measure Unit (SMU). The data acquisition was performed using Labview® software running on an Apple® computer. The base gas used was $1\%O_2/N_2$ which was flown at $3-5 \times 10^{-3}$ L/s over the sample. The mix gases used were 2500ppm Propylene/N_2, 2500 ppm CO/N_2, 2500 ppm Propane/N_2, 1000 ppm H_2/N_2, and 2500 ppm Methane/N_2. The flow rate of the mix gas was adjusted so as to give the appropriate concentration of the sensing gas in the test mixture.

RESULTS AND DISCUSSIONS

The response of the two sensors to varying concentrations of propylene in a base gas of 3.3×10^{-3} L/s (200 cc/min) of $1\%O_2/N_2$ is shown in Figure 3. The initial response of sensor #1 was much lower than that of sensor #2. However, when both sensors were annealed at 1073 K for 3600 s, their responses were almost identical. Moreover, it is seen from Figure 3 that there is no change in the response of sensor #2 to the annealing process. The Pt electrodes of sensor #2 (unlike sensor #1) were cleaned before they were incorporated into the sensor, indicating that the annealing process cleans up the surfaces of the electrodes from

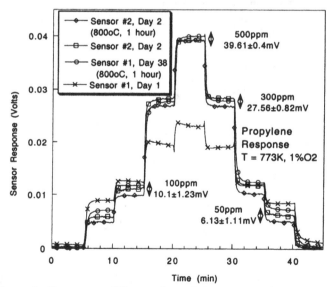

Figure 3. Response of Sensor #1 and #2 to varying concentrations of propylene in a $1\%O_2$ background at T = 773 K.

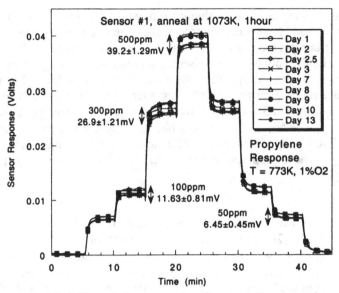

Figure 4. Stability of the response of sensor #1 to varying concentrations of propylene in a 1%O_2 base gas at 773 K.

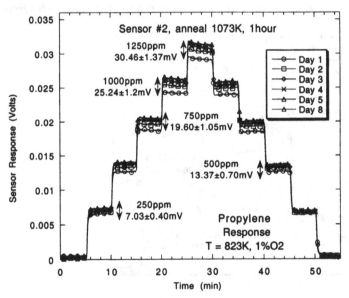

Figure 5. Stability of the response of sensor #2 to varying concentrations of propylene in a 1%O_2 base gas at 823 K.

any adsorbed contaminants. The sensor to sensor reproducibility of these two sensors was within 20% for 50 ppm of propylene and within 1% for 500 ppm of propylene. This reproducibility is attributed to the precise control of the electrode morphology achieved by using a ceramic pellet instead of a sputtered or painted-on gold electrode.[16,17]

Figures 4 and 5 illustrate the stability of the propylene response of sensor #1 and #2 at 773 K and 823 K respectively. The response curves indicate that these sensors are very stable at both these temperatures for up to 10 days. The stability of these sensors is a result of the stable morphology of the ceramic-pellet electrode, which does not coarsen with time. This is in contrast to the stability of the Au electrode (Figure 2) whose morphology does change with time due to the re-crystallization of the Au film.[12]

The response time to 90% of signal level of these sensors is < 25 s at 823 K and < 35 s at 773 K. Although the data points were taken every 2-4 s, only 5% of the data points are shown in the figures for clarity. The response time of these sensors is expected to be a function of the kinetics of the oxidation and reduction reactions (Eqns. 1 & 2) occurring at the two electrodes. However the specific factors that control the kinetics of these reactions are not clearly understood at this time and need to be addressed in detail.

Figure 6 illustrates the response of sensor #2 to various gases at 823 K. In this experiment the base gas (1%O_2/N_2) flow was kept constant at 3.3×10^{-3} L/s (200cc/min) and the mix gas (containing no oxygen) flow was slowly increased from 3.7×10^{-4} L/s to 3.3×10^{-3} L/s. Hence the oxygen partial pressure is not

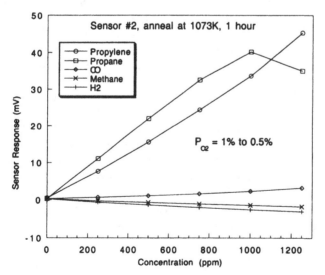

Figure 6. Response of sensor #2 to various gases at 823 K.

constant and varies from 1%O_2 at 0 ppm of mix gas to 0.5%O_2 at 1250 ppm of mix gas. The sensor response has a linear relationship to the concentration of all the gases, while the response to the hydrocarbons (Propylene and Propane) is ≈ 14 times that of CO. Moreover the sensor has only a slightly negative response to H_2 and methane which is probably due to a small mixed-potential on the "pseudo-reference" platinum electrode. This is in contrast to the gold-electrode-based sensor, which has almost no selectivity towards the hydrocarbons (Figure 1).

It is worth mentioning that in Figure 6, the data point for 1250 ppm of propane does not lie on the linear response curve but is actually lower than the response for 1000 ppm of propane. As mentioned earlier, the oxygen concentration was not kept constant in this experiment and for this particular data point was at 0.5%O_2. Hence when 1250 ppm of propane is present alongside 5000 ppm of O_2, we have (see equation 3) crossed over from the lean side to the rich side of stoichiometry.

$$C_3H_8 + 5O_2 \rightarrow 3CO_2 + 4H_2O \qquad (3)$$

Hence these sensors have a linear response curve to the concentration of reducing-gas only in the presence of excess oxygen. Thus these sensors can be operated only on the lean side of stoichiometry and probably would not work for an engine that is running fuel rich.

CONCLUSIONS

The newly fabricated "LSCO/YSZ/Pt" sensor has several desirable characteristics that could be utilized in an on-board hydrocarbon sensor for monitoring automobile tailpipe emissions. The sensor has good selectivity to hydrocarbons and excellent stability and reproducibility. The long-term stability of these sensors and their ability to withstand the hash conditions in an automobile exhaust needs to be examined. However, these can be expected to be reasonably good considering the stability of the individual components used. The YSZ electrolyte is currently being used in the EGO sensor and the platinum electrode should be at least as stable as the platinum used in the catalytic convertor. Moreover, the other electrode (LSCO) is currently being considered as an interconnect material in solid oxide fuel cells and is known to withstand high temperatures (up to 1273 K) and huge swings in the partial pressure of oxygen. However, the response time (< 25 s to 90% of response at 823 K) and sensitivity (≈ 2.5 μV/ppm of propylene at 823 K and 1%O_2) of this device are still major impediments to the practical application of these sensors.

ACKNOWLEDGEMENTS

This work was funded by USCAR DOE CRADA 94-MULT0912-E5-2. Los Alamos National Laboratory is operated by the University of California for the U.S. Department of Energy under contract W-7405-ENG-36.

REFERENCES

[1] J.S. Hepburn, D.A. Dobson, C.P. Hubbard, S.O. Guldberg, E. Thanasiu, W.L. Watkins, B.D. Burns and H.S. Gandi, "A Review of the Dual EGO Sensor Method for OBD-II Catalyst Efficiency Monitoring," *SAE Transactions*, **103** [4] paper 942057 1866-1909 (1994).

[2] W.B. Clemmens, M.A. Sabourin and T. Rao, "Detection of Catalyst Performance Loss Using On-Board Diagnostics," *SAE Transactions*, **99** [4] paper 900062 1866-1909 (1990).

[3] E.M. Logothetis, "ZrO2 Oxygen Sensors in Automotive Applications"; pp. 388-405 in *Advances in Ceramics*, **Volume 3**. Edited by A.H. Heuer and L.W. Hobbs. The American Ceramic Society, Inc. Ohio (1981).

[4] D. E. Williams, P. McGeehin and B. C. Tofield, "Solid Electrolyte Mixed Potential Phenomenon"; pp. 275-278 in *Proc. of the Second European Conf., Solid State Chemistry*. Edited by R. Metselaar, H.J.M. Heijligers and J. Schoonman. Elsevier Publishing Company, Amsterdam (1983).

[5] R. Mukundan, E.L. Brosha, D.R. Brown and F.H. Garzon, "Ceria-Electrolyte-Based Mixed Potential Sensors for the Detection of Hydrocarbons and Carbon Monoxide," *Electrochemical and Solid-State Letters*, **2** [8] 412-414 (1999).

[6] N. Miura, T. Raisen, G. Lu and N. Yamazoe, "Highly Selective CO Sensor Using Stabilized Zirconia and a Couple of Oxide Electrodes," *Sensors and Actuators*, **B47** 84-91 (1998).

[7] R. Sorita and T. Kawano, "A Highly Selective CO Sensor using LaMnO3 Electrode-attached Zirconia Galvanic Cell," *Sensors and Actuators*, **B40** 29-32 (1997).

[8] N. Miura, G. Lu, N. Yamazoe, H. Kurosawa and M. Hasei, "Mixed Potential Type NO_x Sensor Based on Stabilized Zirconia and Oxide Electrode," *J. Electrochem. Soc.*, **143** [2] L33-L35 (1996).

[9] G. Lu, N. Miura and N. Yamazoe, "High Temperature Hydrogen Sensor Based on Stabilized Zirconia and a Metal Oxide Electrode," *Sensors and Actuators*, **B35-36** 130-135 (1996).

[10] T. Hibino, S. Kakimoto and M. Sano, "Non-Nernstian Behavior at Modified Au Electrodes for Hydrocarbon Gas Sensing," *J. Electrochem. Soc.*, **146** [9] 3361-3366 (1999).

[11] F.H Garzon, R. Mukundan and E.L. Brosha, "Mixed Potential Sensors: Theory, Experiments and Challenges," *Solid State Ionics*, **136** 633-638 (2000).

[12] R. Mukundan, E.L. Brosha, D.R. Brown and F.H. Garzon, "A Mixed-Potential Sensor Based on $Ce_{0.8}Gd_{0.2}O_{1.9}$ electrolyte and Platinum and Gold Electrodes," *J. Electrochem. Soc.*, **147** [4] 1583-1588 (2000).

[13] T. Hibino, S. Tanimoto, S. Kakimoto and M. Sano, "High-Temperature Hydrocarbon Sensors Based on a Stabilized Zirconia Electrolyte and Metal Oxide Electrodes," *Electrochemical and Solid-State Letters,* **2** [12] 651-653 (1999).

[14] N. Miura, T. Shiraishi, K. Shimanoe and N. Yamazoe, "Mixed-potential-type propylene sensor based on stabilized zirconia and oxide electrode," *Electrochemistry Communications,* **2** 77-80 (2000).

[15] R. Mukundan, E.L. Brosha and F. Garzon, US Patent, *Serial # 09/770,928*, applied on January 25 2001.

[16] E.L. Brosha, R. Mukundan, D.R. Brown, F.H. Garzon, J.H. Visser, M. Zanini, Z. Zhou and E.M. Logothetis, "CO/HC sensors based on thin films of $LaCoO_3$ and $La_{0.8}Sr_{0.2}CoO_{3-\delta}$ metal oxides," *Sensors and Actuators,* **B69** 171-182 (2000).

[17] R. Mukundan, E.L. Brosha and F. Garzon, US Patent, *Serial # 09/770,359*, applied on January 25 2001.

ZIRCONIA BASED POTENTIOMETRIC NO_x SENSOR UTILIZING Pt AND Au ELECTRODES

D. J. Kubinski, J. H. Visser, R. E. Soltis, M. H. Parsons, K. E. Nietering and S. G. Ejakov
Ford Research Laboratory
MD 3028/ PO Box 2053
Dearborn, MI 48188

ABSTRACT

Mixed-potential sensors based on electrodes with different catalytic activities are simple in design and have been shown to be sensitive to various automotive exhaust gas constituents. Described here is the response of a Pt/YSZ/Au mixed-potential sensor to NO_2 and NO. The sensor response (emf between the Pt and Au electrodes) was measured over the temperature range 500°C-700°C and in accompanying O_2 concentrations $[C(O_2)]$ up to 30%. Also measured were the emf's between each electrode and an air-reference electrode, enabling the responses at the Pt and the Au electrodes to be determined independently. At 600°C and above the sensor response to 150 ppm NO_2 was dominated by the electrochemistry occurring at the Au electrode rather than the Pt. The response time and reproducibility were best at 700°C, although lower temperatures gave larger signals to NO_2. The sensor response to NO_2 decreased considerably as $C(O_2)$ was increased. However, at 500°C and in the presence of 100 ppm NO_2, the emf between the Au and the air-reference electrode exhibited little dependence on $C(O_2)$. At 600°C the sensor showed little sensitivity to 120 ppm NO. Oxidizing some of the NO to NO_2 upstream of the sensor enabled detection of NO levels.

INTRODUCTION

Sensors based on the electrochemical mixed-potential response are known to be sensitive to various automotive exhaust gas constituents. There are reports describing the mixed-potential response of many electrode/electrolyte combinations to gas streams containing H_2,[1] CO,[2-4] hydrocarbons (HCs),[5-7] and NO_x.[8-11] These sensors commonly consist of two electrodes of differing catalytic

---

To the extent authorized under the laws of the United States of America, all copyright interests in this publication are the property of The American Ceramic Society. Any duplication, reproduction, or republication of this publication or any part thereof, without the express written consent of The American Ceramic Society or fee paid to the Copyright Clearance Center, is prohibited.

activities placed on an electrolyte such as yttria-stabilized zirconia (YSZ), an oxygen ion conductor ($O^=$).

Of particular interest for automotive applications are sensors that measure NO_x levels in exhaust. NO_x sensors may be required as part of emission reduction systems for lean-burn and diesel engine applications. The simple design and operation of sensors based on the electrochemical mixed-potential make them an attractive candidate for these applications. A NO_x sensor for automotive applications based on the mixed-potential response is under development by Riken Corporation.[12]

This report describes the response of a Pt/YSZ/Au mixed-potential sensor to concentrations of NO_2 and NO in the presence of up to 30% oxygen in N_2. The Pt and Au were chosen for the catalytic and non-catalytic electrodes since they are easy to fabricate and their response to some of the combustibles (H_2, CO, and C_3H_6) have already been reported.[2,5-7] Although the long-term response of any sensor utilizing Au electrodes at high temperatures can be unstable due to the changing morphology of the Au,[6] this study still elucidates the important issues associated with a mixed-potential based NO_x sensor. Of particular focus in this report is the dependence of the response on temperature and on the concentration of the accompanying O_2. A detailed description of the electrochemical mixed-potential sensing mechanism for NO and NO_2 is given in Ref. 11.

EXPERIMENTAL

In order to determine the separate responses of the Pt and Au electrodes to NO_x, a Pt/YSZ/Au sensor was constructed with a Pt air-reference electrode also placed on the YSZ. The sensor configuration is depicted in Fig. 1. A closed-end tube was made by cementing a YSZ disk (9.5mm-diameter by ~2mm-thick) at the end of a zirconia tube of similar diameter (ID = 7.5mm). The tube was inserted into a quartz tube (~22 mm ID) through which flowed the NO_x test gas at a rate of

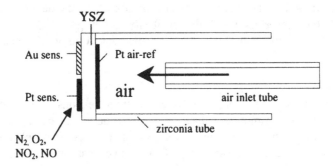

Figure 1. Schematic of sensor configuration. The Pt and Au sensing electrodes are both exposed to the test gas. The Pt reference electrode is exposed to air.

1 l/min. The inside of the closed-end tube was exposed to air instead of the test gas. A ~500 nm-thick sputter deposited Pt film on the surface of the YSZ disk facing the air served as the reference electrode. This electrode was flushed with air at a rate of approximately 0.1 l/min. Separate Pt and Au electrodes were placed on the face of the YSZ disk that was exposed to the NO_x. Both of these electrodes were sputter-deposited, semi-circular in shape and about 500 nm-thick. Electrical contact was made using Pt (or Au) wire, which was bonded to the Pt (Au) electrode with Pt (Au) paste and fired at 900°C. The entire configuration depicted in Fig. 1 was placed inside a tube furnace with measurements in the temperature range 500-700°C. The voltages between the Pt(+) and Au(-) gas-sensing electrodes, between the Pt(+) gas-sensing electrode and the air-reference electrode(-), and between the Au gas-sensing electrode(+) and the air-reference electrode(-) were monitored and are defined as V_{Pt-Au}, V_{Pt} and V_{Au} respectively.

RESULTS AND DISCUSSION
Respsonse to NO_2

Figure 2 shows the sensor response to 0-150 ppm NO_2 at 700°C. The accompanying O_2 concentration [$C(O_2)$] was 0.3% with the balance of the gas N_2. The voltage between the Pt and Au gas sensing electrodes, V_{Pt-Au}, exhibited a non-linear response to NO_2, and was negative in sign. In the presence of a reducing gas such as propylene, V_{Pt-Au} was positive. It is clear that V_{Pt-Au} was dominated by the electrochemistry at the Au electrode. Note that although V_{Pt} changed little upon the addition of 150 ppm NO_2, V_{Au} increased by more than 40 mV. The dependence of V_{Au} on NO_2 concentration is qualitatively similar to the behavior of oxide electrodes reported elsewhere.[8-11]

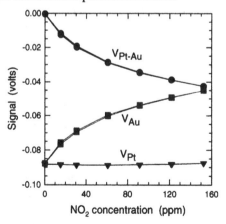

Figure 2. V_{Pt-Au}, V_{Pt} and V_{Au} at T = 700°C as a function of the NO_2 concentration. The O_2 concentration was 0.3%, and the balance of the gas was N_2.

The temperature dependence of the NO_2 sensitivity for $C(O_2) = 0.3\%$ is shown in Figs. 3a-e. These figures plot V_{Pt} and V_{Au} as a function of time for temperatures ranging from 500°C to 700°C. V_{Pt-Au} is not shown in these figures (for clarity), but can be determined from the difference between the two curves. Figure 3f gives the corresponding NO_2 concentration profile as a function of time. The NO_2 was stepped from 0 ppm up to 150 ppm and back down, with 10 minutes spent at each interval. Although V_{Au} was less sensitive to NO_2 at higher temperatures, the signal was faster in response and more reproducible. This is

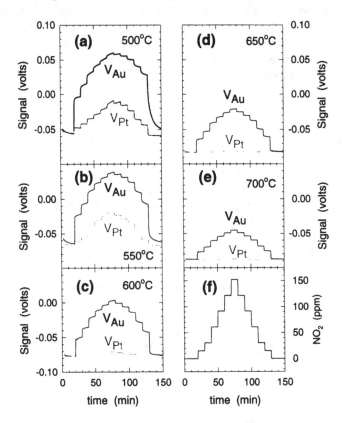

Figures 3a-e. V_{Pt} and V_{Au} as a function of time as the NO_2 concentration was stepped up from 0 ppm to 150 ppm and back down. Shown are data for (a) T = 500°C, (b) T = 550°C, (c) T = 600°C, (d) T = 650°C and (e) T = 700°C. The corresponding NO_2 concentration profile is shown in (f). The time interval of each step of constant NO_2 concentration was ten minutes. The accompanying O_2 concentration was 0.3%, and the balance of the gas was N_2.

especially evident comparing the behavior of V_{Au} at 700°C with that at 500°C. At 500°C V_{Au} did not achieve stability in the ten-minute intervals.

Note that the relative insensitivity of V_{Pt} to NO_2, as demonstrated in Fig. 2 for data at 700°C, was observed in Fig. 3 only for T ≥ 650°C. At lower temperatures V_{Pt} was affected by the NO_2 concentration. For example, at 500°C the increase in V_{Pt} upon the addition of 150 ppm of NO_2 was almost 40% of that for V_{Au}.

Upon examination of Figs. 3a-e it was concluded that operation at 700°C was optimal from the standpoint of response time, stability and reproducibility. Hence, the dependence of the NO_2 response on $C(O_2)$ was studied at this temperature. Figure 4a plots both V_{Pt} and V_{Au} as a function of $C(O_2)$. Shown are data for fixed NO_2 concentrations ranging from 10 ppm to 200 ppm, with $C(O_2)$ up to 20%. Comparison of the differences between the corresponding V_{Pt} and V_{Au} curves confirms the trend of V_{Pt-Au} decreasing in magnitude with higher levels of $C(O_2)$. The expected response to O_2, calculated using the well known Nernst equation: $V = (RT/4F)\ln[C(O_2)/C(O_{2-ref})]$, is also plotted in the figure. The logarithmic dependence of V_{Pt} for $C(O_2) \geq 0.1\%$ is consistent with Nernstian behavior (slopes vary by ~10%) with little influence on the concentration of NO_2. Note that V_{Au} does not vary logarithmically over the $C(O_2)$ range. Each V_{Au} curve for constant NO_2 approached an asymptotic value at low O_2 concentrations. These asymptotic

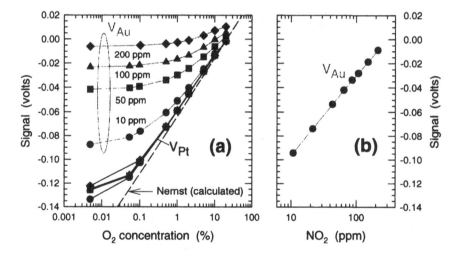

Figure 4. (a) V_{Pt} and V_{Au} at T=700°C as a function of the accompanying O_2 concentration in the range 0.005% to 20%. Shown are data for NO_2 concentrations of 10 ppm (●), 50 ppm (■), 100 ppm (▲), and 200 ppm (♦). The broken line gives the calculated Nernst response to O_2. (b) V_{Au} at T=700°C as a function of the NO_2 concentration with $C(O_2) = 0$.

values were more positive with increasing NO_2. Figure 4b shows that V_{Au} varied logarithmically with NO_2 in the absence of O_2. Similar measurements for V_{Pt} in the absence of O_2 indicated that the deviations from the logarithmic behavior below 0.1% O_2 seen in Fig. 4a are probably a consequence of a small sensitivity of V_{Pt} to NO_2, which is only apparent at 700°C for 200 ppm NO_2 in reduced accompanying O_2 concentrations (< 0.1%).

Figure 5 explores the temperature dependence of the relationship between V_{Au} and $C(O_2)$. Plotted are the responses to 100 ppm NO_2 for T = 500°C, 600°C, 650°C, and 700°C. Consistent with the data in Fig. 3, V_{Au} was found to increase (become more positive) at the lower temperatures. Additionally, the influence of $C(O_2)$ on V_{Au} was reduced at the lower temperatures. At 500°C, 100 ppm NO_2 gave $V_{Au} \approx +65mV$ with minimal dependence on $C(O_2)$. This reduced dependence on $C(O_2)$ is a consequence of $V_{Au} > 0$. In the absence of NO_2, significantly higher values of $C(O_2)$ in the test gas would be required to generate $V_{Au} \approx +65mV$. [It would require $C(O_2)/C(O_{2-ref}) \sim 50$]. Under these conditions, fluctuations in $C(O_2)$ from 0.1%-20% would not be expected to affect V_{Au}. Although the reduced dependence on $C(O_2)$ is ideal for sensor operation, the response times and reproducibility were very poor at 500°C (see Fig. 3a) and would require improvement.

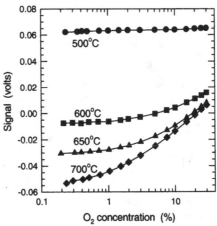

Figure 5. V_{Au} as a function of the accompanying O_2 concentration in the range 0.2% to 30%, with the NO_2 concentration fixed at 100 ppm. Shown are data for $T = 500°C$ (●), $T = 600°C$ (■), $T = 650°C$ (▲) and $T = 700°C$ (◆).

Response to NO

Determination of the response of an electrode to both NO_2 and NO is an important issue since, as demonstrated by Miura et al.,[8-11] the mixed-potential response to each is opposite in sign. This is a consequence of the NO behaving as a reducing gas, and NO_2 as an oxidizing gas. For automotive applications this is problematic since NO_x emissions are typically a mixture of both NO and NO_2.

The sensitivity of the Pt/YSZ/Au sensor at 600°C to NO concentration is shown in Fig. 6 for $C(O_2) = 0.3\%$. In this figure V_{Pt-Au}, V_{Pt} and V_{Au} are plotted as a function of the NO concentration in the range 0-120 ppm. This figure demonstrates the insensitivity of the sensor to NO. However, this result was achieved only after flowing the NO for several hours over the sensor. During this interval a transient signal was observed that was similar to that for a small concentration of NO_2. The transient response was unstable and irreproducible and was not observed with the NO_2 measurements. Its origin is unclear. Additional experiments demonstrated that detection of NO levels in a gas stream could be determined (without transient behavior) by oxidizing the NO into NO_2 upstream of the sensor. For example, a Pt foil catalyst at 410°C and placed upstream of the sensor was estimated to oxidize about 20% of the NO to NO_2.

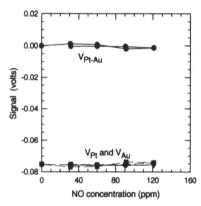

Figure 6. V_{Pt-Au}, V_{Pt} and V_{Au} at T = 600°C as a function of the NO concentration. The O_2 concentration was 0.3%, and the balance of the gas was N_2.

CONCLUSIONS

This report demonstrated that a Pt/YSZ/Au mixed-potential sensor was quite sensitive to low levels of NO_2 and insensitive to NO. At 600°C and above the response to NO_2 was dominated by the electrochemistry occurring on the Au electrode rather than the Pt. Lower temperatures gave a larger response to NO_2, but the response time and reproducibility were best at 700°C, with 15 ppm NO_2 easily detected. The response to NO_2 was very dependent on $C(O_2)$. However at 500°C and in 100 ppm NO_2, the emf between the Au electrode and the air-reference electrode, V_{Au}, was positive in sign and exhibited a little dependency on $C(O_2)$. Improvement of the response time for V_{Au} may enable the possibility of a mixed-potential based NO_2 sensor with reduced dependency on $C(O_2)$.

ACKNOWLEDGEMENT
The author's thank Mr. T. Lockwood for his valuable assistance with the data acquisition system.

REFERENCES

[1] G. Lu, N. Miura and N. Yamazoe, "*High-temperature hydrogen sensor based on stabilized zirconia and a metal oxide electrode*," Sens. and Actuators B, **35-36**, 130 (1996).

[2] D. E. Williams, P. McGeehin and B. C. Tofield, "*Solid electrolyte mixed-potential phenomena*," Proc. Second European Conf., Solid State Chemistry, Veldhoven, The Netherlands, June 7-9, 275 (1982).

[3] R. Sorita and T. Kawano, "*A highly selective CO sensor using $LaMnO_3$ electrode-attached zirconia galvanic cell*," Sens. and Actuators B, **40**, 29 (1997).

[4] N. Miura, T. Raisen, G. Lu and N. Yamazoe, "*Highly selective CO sensor using stabilized zirconia and a couple of oxide electrodes*," Sensors and Actuators B, **47**, 84 (1998).

[5] T. Hibino, S. Wang, Y. Kuwahara, S. Kakimoto and M. Sano, "*Detection of propylene under oxidizing conditions using zirconia-based potentiometric sensor*," Sens. and Actuators B, **50**, 149 (1998).

[6] R. Mukundan, E. L. Brosha, D. R. Brown and F. H Garzon, "*Ceria-electrolyte-based mixed-potential sensors for the detection of hydrocarbons and carbon monoxide*," Electrochemical and Solid State Letters, **2**, 412 (1999).

[7] R. Mukundan, E. L. Brosha, D. R. Brown and F. H Garzon, "*A mixed-potential sensor based on a $Ce_{0.8}Gd_{0.2}O_{1.9}$ electrolyte and platinum and gold electrodes*," J. Electrochem. Soc., **147**, 1583 (2000).

[8] N. Miura, H. Kurosawa, M. Hasei, G. Lu and N. Yamazoe, "*Stabilized zirconia-based sensor using oxide electrode for detection of NOx in high temperature combustion-exhausts*," Solid State Ionics, **86-88**, 1069 (1996).

[9] G. Lu, N. Miura and N. Yamazoe, "*High temperature sensors for NO and NO_2 based on stabilized zirconia and spinel-type oxide electrodes*," J. Mater. Chem., **7**, 1445 (1997).

[10] N. Miura, G. Lu and N. Yamazoe, "*High-temperature potentiometric/amperometric NOx sensors combining stabilized zirconia with mixed-metal oxide electrode*," Sens. and Actuators B, **52**, 169 (1998).

[11] G. Lu, N. Miura and N. Yamazoe, "*High-temperature NO and NO_2 sensor using stabilized zirconia and tungsten oxide electrode*," Ionics, **4**, 16 (1998).

[12] T. Ono, M. Hasei, A. Kunimoto, T. Yamamoto and A. Noda, "*Performance of the NOx sensor based on mixed potential for automobiles in exhaust gases*," JSAE Review, **22**, 49 (2001).

PACKAGING PLANAR EXHAUST SENSORS FOR HOSTILE EXHAUST ENVIRONMENTS

C. Scott Nelson
Delphi Automotive Systems
1601 N. Averill Ave.
MC 485-220-010
Flint, MI 48556 USA

ABSTRACT
An automotive exhaust system presents a very challenging environment for chemical sensors. These sensors must monitor or measure a gas species, usually throughout the life of the vehicle. They can be subjected to severe cold (-40°C) and excessive temperatures reaching (> 1000°C), extreme vibrations, stone impacts, and a constant stream of potential poisons. Survival of these conditions is a requirement, while only minimal sensor degradation is allowed over the sensor life. Through finite element methods, a planar sensor package has been designed to be robust to this severe environment while meeting new customer requirements for overall shorter profiles.

INTRODUCTION
Exhaust sensors, and in particular Oxygen sensors, have been used in automotive exhaust systems for over 25 years. During this time, requirements for sensor packaging have mandated the sensor be extremely robust under sometimes very hostile conditions.

The package must protect the sensor from chemical attack, force, impacts, vibration, and heat for a minimum of 160,000 km. Chemical attack comes externally from road salts, antifreeze, water, soda etc. Internal chemicals come from contaminants in the fuel, materials from gaskets, engine oil, and antifreeze. Forces on the sensor can be from pulling on the wires, or a point load on the body. Impacts can come from dropping the sensor or stones impacting the body. Several sources of vibration can cause a large number of potential resonant frequencies; these sources include road noise, the engine, and accessories. Both exhaust and ambient temperatures are important. Internally, the sensor can be exposed to exhaust temperatures in excess of 1000°C; ambient temperatures have been known to reach 300°C with no air flow.

Protecting the sensor from external chemical attack typically involves using corrosion resistant stainless steels and hermetically sealing the interior of the sensor from the environment. Further enhancement involves the sensor element using a sealed air reference. The sensor provides its own electrochemical oxygen "pump" in both rich and lean exhaust mixtures, preventing the possibility of air reference contamination and allowing for a simplified package. Protecting from chemical attack from the exhaust stream involves shielding optimization and element protective coatings.

Protecting against force, impacts and vibration can be sensor dependant based on the sensor packaging's "weakest link". However, one general trend tends to follow: the shorter the sensor, the less prone a sensor is to damage caused by vibration, point forces, and impacts - due to the cantilever effect. This of course is directionally incorrect for protecting the sensor against heat - the shorter the sensor, the higher the temperature.

This paper will discuss several thermal optimization strategies for lowering the temperature in critical areas of a sensor in order to reduce the overall height of the package. Examples of these strategies will be put into practice and evaluated using finite element analysis (FEA). This will be followed by a comparison of the FEA results and actual test data.

ANALYSIS

In determining the package vertical dimensions for a sensor, typically the most important parameter is the maximum steady state temperature to which the sensor will be exposed. In exhaust sensors there are usually two critical areas to observe when determining the height of the sensor. First, all materials should be below their maximum use temperatures when the sensor is exposed to its steady state maximum temperatures. The component that is most sensitive is usually the wire seal at the top of the sensor. This material is usually made of some type of elastomer in order to seal around the wires; if the material reaches temperatures in excess of its operating temperature, it will begin to degrade and can cause contaminants to enter the sensor. The second area of concern is the terminal contact interface. Depending on the material, and type of interface, terminal materials can greatly increase their susceptibility to oxidation with increasing temperature. An additional complication is increased polymerization due to the elastomer seal out-gassing due to over temperature exposure. Thus the two areas of concentration in optimizing the sensor packaging is to reduce the temperatures in the area of the element-terminal connection and the wire seal.

The primary mode of heat transfer inside a sensor package is through conduction:

$$Q = \frac{Ak}{L}(T_1 - T_2) \qquad (1)$$

Where Q is the heat energy transferred, A is the cross sectional area perpendicular to the direction of heat flow, k is the thermal conductivity of the material, L is the length of the heat path, T_1 is the starting temperature, and T_2 is the temperature at the distance L.

Although equation (1) is rather simple, it is seldom fully optimized. Often the area A is chosen based on previous sensor designs, and the thermal conductivity k is based on previous materials used, thus the length L in equation (1) is used to reduce the heat energy that reaches the critical components. Equation (1) can also be used to maximize heat transfer away from a component.

In sensor design, there are three basic strategies used for minimizing temperatures at a specific component (given a certain set of boundary conditions): restrict heat flow up the sensor, promote radial heat flow to the ambient environment, and increase heat flow through a component.

Each of the above mentioned thermal strategies were evaluated and optimized using an iterative approach. Materials, areas, heights, geometry, and component position were modified both separately and in combination in order to arrive at the final design.

FINITE ELEMENT ANALYSIS

Algor steady state thermal modeling software was used to optimize the sensor packaging. The model took into consideration conduction, convection and radiation. Since the analysis was evaluated at maximum temperature conditions, the temperature of the sensing element was greater than the normal operating temperature and thus the heater was inactive and did not need to be taken into consideration.

Although the sensor is not truly symmetrical, in order to simplify the analysis, the sensor was represented as a 2-D axisymmetric model. In situations where a component was not symmetrical, an equivalent area was created so that when it was swept 360 degrees, both the model and actual component would have the same volumes.

Temperature boundary conditions used in the model were taken from worst case vehicle conditions. Convection coefficients were taken from Chen, et. al. (1) for wide open throttle conditions.

RESULTS AND DISCUSSION

Figure 1. shows the optimized sensor. Figure 2 shows the FEA model temperature results.

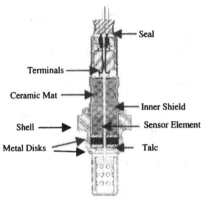

Figure 1 Optimized OSL Sensor

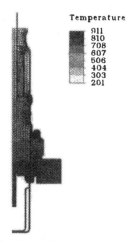

Figure 2. Optimized sensor temperature results

Table I., shows the target temperatures, FEA temperatures from the initial design, the final design, and temperatures as measured with thermocouples in the final design.

Table I. Target, FEA, and actual test results

	Terminal Temp (C)	Seal Temp (C)
Target	350	250
Initial Sensor (FEA)	435	324
Final Sensor (FEA)	345	228
Test Results	349	220

The approach taken was to start modeling the sensor at the target height, and iterate until the desired height, temperature constraints and manufacturability was achieved. As shown, the actual test data is within 4% of the modeled FEA sensor and slightly below the target temperatures (the seal temperature is significantly lower than the target temperature due to the convection coefficients being higher on the test fixture than indicated in the target).

As previously mentioned, the environmental conditions around the sensor are critical. The previous results were modeled using free convection only. If a 0.75 m/s (~1.7 mph) airflow is allowed to flow across the sensor upper shielding, the terminal temperature is reduced to 338°C and the seal temperature is 249°C. If the airflow is increased to 1.8 m/s (4 mph), the temperatures are reduced to 323°C and 210°C for the terminal and seal respectively. Table II, shows a comparison of the same sensor configuration, with different airflows across the upper shield.

Table II: FEA results with various airflows

FEA Model	Terminal Temperature (°C)	Seal Temperature (°C)	Hex Temperature (°C)
Thermally optimized sensor – free convection	352	256	687
Thermally optimized sensor – 0.75 m/s (1.7 mph) airflow	338	249	674
Thermally optimized sensor – 1.8 m/s (4 mph) airflow	323	210	672

Comparing the airflows in Tables II, it is obvious the role that airflow plays in sensor temperatures. Even though each row of results use ambient air temperatures of 150°C, the critical temperatures decrease considerably with only a slight breeze since the convection coefficient is greatly influenced by the airflow. This comparison shows how critical it is to package an exhaust sensor so that it has proper airflow.

The shell temperature, also called the hex temperature, is often used to make reference to the exhaust temperature. This can often be misleading when comparing different designs. The thermal conductivity and heat path of the overall package influences the hex temperature. This can be seen from the summary results as shown in Table II. As shown, even though the exhaust temperature remains the same for each iteration, the hex temperature varies according to the design.

SUMMARY

It has been shown that FEA analysis is very useful in thermally optimizing a sensor. Three main temperature reducing techniques were used: restricting temperature up the sensor (low thermal conductivity materials, reducing cross sectional area, utilizing air gaps), promoting radial conduction (eliminate transverse air gaps, solid material contact), and allowing heat to pass through a critical component (increasing thermal conductivity above the critical component). From the starting sensor configuration, the temperatures were reduced by 83°C at the terminal contacts and 68°C at the seal. Comparison of FEA modeling and actual data shows that they are within 4%. Non-traditional sensor components may be necessary to achieve lower profile sensors that retain the same temperature requirements as their longer counterparts. It has also been shown how critical airflow around the outside of the sensor is in minimizing temperatures. It is important to model the worst case conditions of a sensor in order to prevent out gassing of the seal and oxidation at the terminals. Although thermal optimization is only one step in making a durable sensor, it plays a major role in contributing to a world class design. The thermally optimized sensor is currently undergoing an extensive battery of tests, and is showing that it is an extremely durable design.

REFERENCES

[1] David Chen, et al., "Optimization of Oxygen Sensor" (SAE 2000-01-1364).

IMPORTANCE OF GAS DIFFUSION IN SEMICONDUCTOR GAS SENSOR

Noboru Yamazoe, Go Sakai, Naoki Matsunaga,
Nam-Seok Baik and Kengo Shimanoe
Department of Molecular and Material Sciences,
Interdisciplinary Graduate School of Engineering Sciences,
Kyushu University,
Kasuga-shi, Fukuoka 816-8580, Japan

ABSTRACT
The influences of gas transport on the gas sensing properties of thin film semiconductor gas sensor are discussed theoretically, based on the diffusion equation assuming Knudsen diffusion and first order surface reaction of a target gas (H_2). By solving the diffusion equation under steady state conditions and introducing some simplifying assumptions, it is possible to formulate the response (sensitivity, S) of a thin film as a function of film thickness (L) and temperature (T). The observed dependence of S on L as well as the well known volcano shaped correlation between S and T can be simulated well theoretically. The solution of the diffusion equation under non-steady state conditions discloses transient phenomena in gas concentration profile and response.

INTRODUCTION
A semiconductor gas sensor detects an inflammable gas (target gas) in air from a decrease in electrical resistance. The gas sensitive resistor (element) of it, fabricated in the form of a block, thick film or thin film, is basically a porously packed assembly of nano particles of an n-type semiconducting oxide like SnO_2. Through the elucidation of the role of adsorbed oxygen on the oxide surface, the effects of oxide grain size, and so on, it has been widely accepted that the target gas reacts with the adsorbed oxygen and that the resulting decrease of the adsorbed oxygen concentration leads to decreases in Schottky barriers at the grain boundaries and so in the electrical resistance[1-3]. These concepts, though useful for understanding the gas sensing mechanism, are not always sufficient to explain the gas sensing properties of the element as a whole. What is missing is transport of the target gas inside the element. When the element is exposed to the target gas in air, the gas molecules diffuse in the element while reacting progressively with the adsorbed oxygen. The

target gas concentration inside should be smaller than that in the outside and decrease with increasing diffusion depth. The gas concentration profile inside the element will be dependent on the relative rates of diffusion and reaction and this will depend further on the microstructure of the element.

Such gas transport problems have been studied well in the field of heterogeneous catalysis, where reactant gas molecules diffuse in porous solid catalysts. The importance of gas transport has also been recognized in the semiconductor gas sensor[4-7]. For example, the response (sensitivity) to a target gas of a fixed concentration goes though a maximum on increasing operating temperature. This phenomenon has been understood as follows[8]. In the lower temperature region, the gas molecules having diffused in the element react with the adsorbed oxygen more actively with increasing temperature (ascending part of the response). In the higher temperature region where the reactivity becomes too large, the gas molecules tend to be consumed up before diffusing deep inside the element (descending part). The temperature at the maximum shifts downward when a catalytically active material like PdO is added to the element[9]. The location and geometry of electrodes affect the gas sensing properties of thick film elements[10]. These phenomena are also associated with gas transport.

Despite such recognition, however, the relation between gas sensing properties and gas transport properties has been left unanalyzed. Unlike the heterogenous catalysis, the semiconductor gas sensor concerns the electrical resistance of the whole element, making analysis far more complicated. Even under steady state, the resistivity would change from site to site in the element due to the gas transport problem. One should know the structure and gas concentration at every site of the element. For this purpose, one needs to have the sensor element well defined in microstructure, typically in grain size, pore size and thickness.

Recently we obtained fairly well defined thin films of SnO_2 by a wet method, i.e., spin-coating from a hydrothermally treated sol suspension of SnO_2. The SnO_2 grains coated were fairly stable to thermal growth at elevated temperature, allowing very stable gas sensing properties[11,12]. Encouraged with this finding, we tried to analyze the relevance of gas transport properties to the gas sensing properties based on these films[13]. The present paper aims at summarizing briefly the results of steady state analysis[14,15] as well as nonsteady state analysis[16].

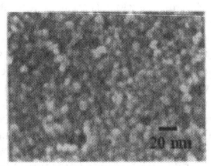

Figure 1. FE-SEM image of SnO_2 grains in thin film calcined at 600 °C.

Sol SUSPENSION DERIVED THIN FILMS OF SnO_2

For the analysis of gas transport related properties, the thin films prepared by spincoating from a hydrothermally treated sol suspension of SnO_2 (SnO_2 content 1.8 mass %) were fairly satisfactory[12]. The morphology of a typical thin film, 100 nm thick, spin-coated on an alumina substrate and calcined at 600 °C for 5 h, is shown in figure 1. Tiny grains of SnO_2 with a mean size of 6 nm in diameter were stacked into a uniform layer. Crack-free thin films with thicknesses up to about 300 nm could be prepared by repeating the spin-coating.

The H_2 sensing properties of such a thin film are illustrated in figure 2. The response (or normalized conductance), R_a/R_g, where R_a and R_g are the resistances of the film in air and the sample gas, respectively, increased linearly with the H_2 concentration in air (5 - 4000 ppm) in the logarithmic scale. It is noted that the response to 5ppm H_2 was as high as about 40, confirming the high sensitive nature of the film. Also notably, the slope of the linear correlation was 0.97, indicating almost linear dependence of R_a/R_g on the H_2 concentration. A remarkable feature of the sol suspension derived thin films could be found in the stability of sensing properties. As an example, the result of a short-term stability test for a freshly prepared thin film (with no aging treatments) is shown in figure 3. To accelerate any possible changes, both the air and the sample gas (800ppm H_2 in air) were humidi-

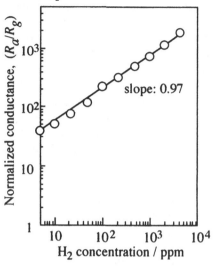

Figure 2. Normalized conductance of a thin-film sensor as a function of H_2 concentration (350 °C).

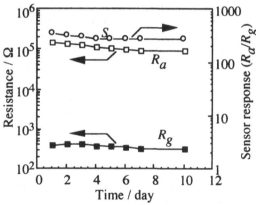

Figure 3. Behavior of electrical properties of the thin film during continuous sensor operation under humidified conditions at 500 °C.
Ra: resistance in air (RH 50%)
Rg: resistance in 800 ppm H2 in air (RH 50%)
S: sensor response (Ra/Rg)

fied to relative humidity 50% at room temperature, while the operating temperature (500 °C) was set to be higher than in usual cases. The resistance in air (R_a) moved down only slightly before being stabilized almost completely in a week. The resistance in the sample gas (R_g), which was measured once a day, tended to be even more stable. The response (R_a/R_g) was also fairly stable, as shown. Such high stability seems to reflect the thermal stability of the hydrothermally treated SnO_2 grains involved.

The influence of film thickness was investigated by using such thin films of various thicknesses (80-300 nm). As shown by open circles in figure 4, the response (sensitivity) to 800 ppm H_2 in air (dry) was found to decrease with increasing film thickness. With the other conditions being fixed, this behavior is likely ascribable to gas-transport.

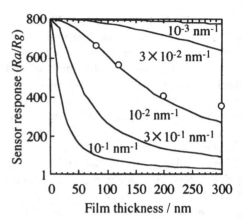

Figure 4. The response (Ra/Rg) to 800 ppm H2 vs film thickness correlations. ○: observed, -correlation lines simulated based on eq (4).

RESPONSE DEPENDENCE ON THICKNESS

It is known that gas molecules moves inside porous materials by molecular diffusion, Knudsen diffusion or surface diffusion, depending on the size of the pores included, and that Knudsen diffusion dominates in the pores of 1-100 nm in diameter (mesopores). The diffusion of this type is assumed to take place in the SnO_2 thin films just mentioned, while the gas (H_2) molecules are assumed to react with the adsorbed oxygen of the SnO_2 grains in the first order reaction kinetics. Under these assumptions, a diffusion equation is formulated as follows.

$$\frac{\partial C}{\partial t} = D_K \frac{\partial^2 C}{\partial x^2} - k \cdot C \qquad (1)$$

Here, C is target gas concentration, t time, x depth from the film surface, D_K Knudsen diffusion coefficient and k rate constant. Under steady state ($\partial C/\partial t=0$), the following boundary conditions are set based on a model shown in figure 5 (a) : $C = C_s$ (gas concentration outside) at $x = 0$ (surface) and $dC/dx = 0$ at $x = L$ (bottom). The solution of eq (1) is expressed as follows.

$$C = C_s \cdot \cosh((L-x)\sqrt{k/D_K})/\cosh(L\sqrt{k/D_K}) \qquad (2)$$

Gas concentration profiles inside a thin film of $L=300$ nm under several conditions are shown in figure 4. C decreases more sharply on going inside as $\sqrt{k/D_K}$ increases:

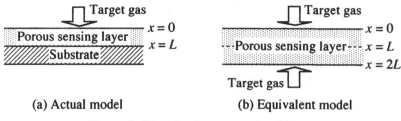

(a) Actual model   (b) Equivalent model

Figure 5. Models of a gas sensing film.

The gas penetration depth becomes limited in the surface region when $\sqrt{k/D_K}$ is large. It is remarked that an equivalent model is available as depicted in figure 5(b). The thickness is doubled here with symmetric boundary conditions, $C=C_s$ at $x=0$ and $2L$. This model gives rise to the same solution (2), and the solution up to $x=L$ can be utilized for the analysis. It follows that the reflection of gas molecules at the bottom in the actual model is equivalent to the diffusion flux (at $x=L$) of the gas molecules which have diffused from the other surface in the equivalent model.

Once the gas concentration profile is known, it is rather easy to formulate the response (sensitivity) of the film to the target gas. The film is supposed to consist of infinitesimally thin horizontal sheets. The conductance of a sheet at depth x, $\sigma(x)$, is assumed to be linear to $C(x)$.

$$\sigma(x)/\sigma_0 = 1 + a \cdot C(x) \qquad (3)$$

Here σ_0 is the conductance in air. The proportionality constant, a, defined here as sensitivity coefficient, expresses the sensitivity of the sheet to the gas. The linearity assumed here is based on the linear response to the gas concentration shown in figure 2. Expression of $\sigma(x)$ in non linear cases have been discussed by Williams[1]. The conductance of the whole film in air ($1/R_a$) and in the sample gas ($1/R_g$) are obtained by integrating σ_0 and $\sigma(x)$ over the whole thickness, respectively. By using eq (3) for $\sigma(x)$, the response (sensitivity) of the film is derived to be as follows.

$$S = \frac{R_a}{R_g} = 1 + \frac{a \cdot C_s}{m} \tanh m \qquad (4)$$

$$m = L\sqrt{k/D_K} \qquad (5)$$

S is uniquely determined by m if a and C_s are fixed, and so by thickness L for given $\sqrt{k/D_K}$. The S vs L correlations are depicted for various values of $\sqrt{k/D_K}$ in figure 4. Here C_s = 800 ppm and a is assumed to be 1 ppm^{-1}. The observed data lie fairly close to the correlation line for $\sqrt{k/D_K}$ =10^{-2} nm^{-1}, supporting that the variations in response are caused by gas transport as suspected.

RESPONSE DEPENDENCE ON TEMPERATURE

Eq (4) includes temperature dependent coefficients or constants, i.e., D_K, k, a and C_s. If the temperature dependence of these terms are expressed, the equation would predict the temperature dependence of S. The dependence of D_K and C_s are given by

$$D_K = 4r/3 \cdot \sqrt{2RT/\pi M} \quad (6)$$

$$C_s = (293/T) \cdot C_m \quad (7)$$

Here r is pore radius, M molecular weight of the target gas, and C_m nominal concentration of the target gas in the sample gas mixed at 293 K, and R and T have their usual miauings. The dependence of k would be expressed as follows,

$$k = k_0 \exp(-E_k/RT) \quad (8)$$

where k_0 is constant and E_k is activation energy of the surface reaction. The dependence of sensitivity coefficient, a, is not certain yet, but, as a is associated with the relative change of adsorbed oxygen concentration on the oxide grain due to the surface reaction as well as with the resultant transduction into a resistance change, the dependence of a is assumed to be also of Arrhenius type as follows.

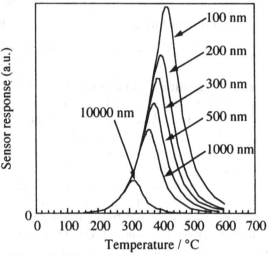

Figure 6. The dependence of sensor response on temperature at various film thicknesses, simulated under the conditions; a_0=139400 ppm^{-1}, A=1.0*10^{10} nm^{-1}·K$^{-1/4}$, Ea=58.2 kJ·mol^{-1}, Ek=299.3 kJ·mol^{-1}.

$$a = a_0 \exp(-E_a/RT) \quad (9)$$

Here a_0 is constant and E_a is apparent activation energy of the signal transduction. Insertion of Eqs (6) through (9) into Eq (5) derives the following equation

$$S = 1 + \frac{a_0 \cdot 293 \cdot C_m}{A \cdot L} \cdot T^{-\frac{3}{4}} \cdot \exp\left(-\frac{2E_a - E_k}{2RT}\right) \cdot \tanh\left(A \cdot L \cdot T^{-\frac{1}{4}} \cdot \exp\left(-\frac{E_k}{2RT}\right)\right) \quad (10)$$

$$A = (3k_0/4r)^{1/2} \cdot (\pi M/2R)^{1/4} \quad (11)$$

Roughly speaking, the temperature dependence of S is determined mainly by the $\exp\{-(2E_a-E_k)/2RT\}$ term, and S goes thorough a maxim when $2E_a<E_k$. The tempera-

ture dependence of S is illustrated in figure 6, where the constants are set as indicated. It is informative that the familiar volcano shaped correlation between S and T can be generated by simulation in this way. The ascent is attributed to an increase in a with increasing temperature while the descent is to a decrease in gas penetration depth inside the film. The volcano is seen to move to the higher temperature side, accompanied by an increase in its height, as L decreases. This tendency reflects that the gas transport problem becomes less important as L decreases.

5. TRANSIENT PROPERTIES

So far, we were concerned with the gas sensing properties at steady state. It is possible to solve the diffusion equation (1) under non steady state conditions based on the equivalent model in figure 5 (b). The boundary conditions are the same as those used previously: $C=C_s$ at $x=0$ and $2L$. The initial condition is: $C=0$ at any points inside the film at $t=0$. The solutions is expressed as follows.

$$C = C_s \left[1 - \frac{2}{\pi} \sum_{n=1}^{\infty} \frac{1-(-1)^n}{n^2} \cdot \left\{ \exp(-\omega_n^2 t) + \frac{k\{1-(1+\omega_n^2 t)\cdot \exp(-\omega_n^2 t)\}}{\omega_n^2 + k\{1-\exp(-\omega_n^2 t)\}} \right\} \cdot \sin\frac{n\pi}{2L}x \right]$$

$\omega_n = \pi D_K^{1/2} / 2L$ (12)

This equation consists of an infinity of terms, but the contribution of each term decreases with going to higher terms, so that summing up the initial 10,000 terms is sufficient to obtain a gas concentration profile at any selected value of t. An important feature of eq (10) is that D_K and k affects the concentration profile independently, unlike the case of the steady state analysis. Figure 7 shows gas concentration profiles at various times after exposure to the target gas (1,000 ppm H_2) under the given set of constants as indicated. As t increases, the profile develops from a) through e). Very remarkably, the gas concentration at points close to the bottom (x= 1,000 nm) goes up to an excessively high level, as illustrated by c) and d), before going down to a stationary level e). The profile e) is confirmed to co-

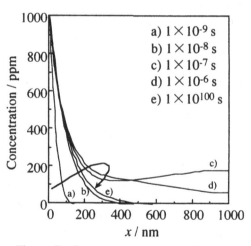

a) 1×10^{-9} s
b) 1×10^{-8} s
c) 1×10^{-7} s
d) 1×10^{-6} s
e) 1×10^{100} s

Figure 7. Gas concentration profiles inside the film at various periods of time after exposure to 1000 ppm H2. $D_K=10^{12}$ nm^2·s^{-1}, $k=10^8$ s^{-1}, $L=1000$ nm

incide precisely with the stationary one shown by eq (2). This transient phenomenon takes place because the gas molecules are reflected by the substrate.

The transient of the response (S) can be obtained in the same way as before, which is expressed as follows.

$$S = \frac{R_a}{R_g} = 1 + a \cdot C_s \left[1 - \frac{4}{\pi^2} \sum_{n=1}^{\infty} \frac{1-(-1)^n}{n^2} \cdot \left\{ \exp(-\omega_n^2 t) + \frac{k\{1-(1+\omega_n^2 t) \cdot \exp(-\omega_n^2 t)\}}{\omega_n^2 + k\{1 - \exp(-\omega_n^2 t)\}} \right\} \right]$$

(13)

Figure 8 illustrates the response to H_2 as a function of time. The conditions are the same as those in figure 7 except for k which changes between 10^5 to 10^9 s^{-1}. Reflecting the transient in gas concentration profile just mentioned, the response goes through a maximum, and the maximum becomes more conspicuous as k increases. Nevertheless, the time for the maximum is dependent of k, because it is given approximately by $4L^2/\pi^2 D_K$. The transient finishes in about a microsecond under the present conditions, but it may be made observable experimentally under some particular conditions of D_K and L.

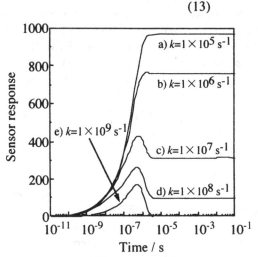

Figure 8. Simulated time course of sensor response to 1000 ppm H2 for various values of k.

EXTENTION TO THICK FILMS AND SINTERED BLOCKS

Compared to the thin films so far discussed, thick film- and sintered block-type elements are more complicated in microstructure, with the formation of secondary particles (1-10 μm in diameter) or higher-order structure. In addition to the mesopores, macropores (pore radius above 100 nm) are formed among the secondary particles. It is known that molecular diffusion takes place in macropores, which is usually far quicker than Knudsen diffusion. This difference in diffusion mechanism allows to estimate the gas sensing behavior of these elements qualitatively. For simplicity, the element is assumed to consist of uniform secondary particles which are contacting to each other as schematically shown in figure 9. Since the gas concentration in the macropores can be assumed to be close to that outside the element in the steady state, the gas concentration gradient develops mainly inside the secondary particles.

It is easily understood that, under usual operating conditions, the gas penetration depth (gas sensitive region) is limited within a rather shallow layer beneath the periphery of each secondary particle. The situation of gas-solid interaction in the secondary particles is rather similar to that of a thin film with large L. It follows that the sensitivity would decrease as the secondary particle size becomes larger, while a volcano shaped S vs. T correlation would result similarly from the gas transport problem inside the secondary particles.

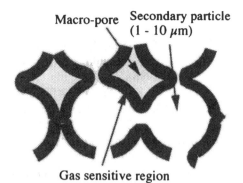

Figure 9. Schematic drawing for microstructure of thick film and block-type elements.

CONCLUSIONS

The dependence of the response (sensitivity) of an SnO_2 thin film element on film thickness and operating temperature in the steady state can be formulated by considering the diffusion and surface reaction of gas molecules inside the thin film. On sudden exposure of the film to the target gas, transient phenomena take place for gas concentration profile and gas response as a result of the competition between diffusion and surface reaction. These results can be applicable, to a limited degree, to a thick film- or sintered block-type element.

REFERENCES

[1] N. Yamazoe, J. Fuchigami, M. Kishikawa, and T. Seiyama, "Interactions of Tin Oxide Surface with O_2, H_2O, and H_2," *Surface Science*, **86**, 335-344 (1979).

[2] K. Ihokura, "Tin Oxide Gas Sensor for Deoxidizing Gas," pp.43-50 in *New Materials & New Processes In Electrochemical Technology* Vol. 1, Edited by M. Nagayama et al., JEC Press Inc., 1981.

[3] C. Xu, J. Tamaki, N. Miura, and N. Yamazoe, "Grain size Effects on Gas Sensitivity of Porous SnO_2-based Elements," *Sensors and Actuators B*, **3**, 147-155 (1991).

[4] D.E. Williams, "Conduction and Gas Response of Semiconductor Gas Sensors,"; pp. 71-123 in *Solid State Gas Sensors*, Edited by P. T. Moseley and B. C. Tofield. Adam Hilger, Bistrol and Philadelphia, 1987.

[5] Y. Shimizu, T. Maekawa, Y. Nakamura, M. Egashira, "Effects of gas diffusivity and reactivity on sensing properties of thick film SnO2-based sensors," *Sensors and Actuators B*, **46**, 163-168 (1998).

[6] D.E. Williams, G.S. Henshaw, K.F.E. Pratt and R. Peat, "Reaction-Diffusion Effects and Systematic Design of Gas-sensitive Resistors based on Semiconducting

Oxides," *Journal of Chemical Society Faraday Transduction*, **91**[23], 4299-4307 (1995).

[7]X. Vilanova, E. Llobet, R. Alcubilla, J.E. Sueiras, and X. Correig, "Analysis of the conductance transient in thick-film tin oxide gas sensors," *Sensors and Actuators B*, **31**, 175-180 (1996).

[8]N. Yamazoe, Y. Kurokawa, and T. Seiyama, "Effects of additives on semiconductor gas sensors," *Sensors and Actuators*, **4**, 283-289 (1983)

[9].T. Seiyama, H. Furuta, F. Era, and N. Yamazoe, "Gas detection by activated semiconductive sensor," *Denki Kagaku*, **40**, 244-249 (1972).

[10]U. Jain, A.H. Harker, A.M. Stoneham, and D.E. Williams, "Effect of Electrode Geometry on Sensor Response," *Sensors and Actuators B*, **2**, 111-114 (1990).

[11]N.S. Baik, G. Sakai, N. Miura, N. Yamazoe, "Preparation of Stabilized Nanosized Tin Oxide Particles by Hydrothermal Treatment," *Journal of American Ceramic Society*, **83**[12] 2983-2987 (2000).

[12]N.S. Baik, G. Sakai, K. Shimanoe, N. Miura, N. Yamazoe, "Hydrothermal treatment of tin oxide sol solution for preparation of thin-film sensor with enhanced thermal stability and gas sensitivity," *Sensors and Actuators B*, **65**, 97-100 (2000).

[13]G. Sakai, N. S. Baik, N. Miura, N. Yamazoe, "Gas sensing properties of tin oxide thin films fabricated from hydrothermally treated nanoparticles –Dependence of CO and H_2 response on film thickness-," *Sensors and Actuators B*, **77**, 116-121 (2001).

[14]G. Sakai, N. Matsunaga, K. Shimanoe, N. Yamazoe, "Diffusion Equation-based Study of Thin Film Type Semiconductor Gas Sensor (1) Sensitivity Behavior under Steady State," *Chemical Sensors*, Vol. 17, Supplement A, 1-3 (2001).

[15]G. Sakai, N. Matsunaga, K. Shimanoe, N. Yamazoe, "Theory of Gas-diffusion Controlled Sensitivity for Thin Film Semiconductor Gas Sensor," *Sensors and Actuators B*, **80**, 125-131 (2001).

[16]N. Matsunaga, G. Sakai, K. Shimanoe, N. Yamazoe, "Diffusion Equation-based Study of Thin Film Semiconductor Gas Sensor (2) Response Transient," *Chemical Sensors*, Vol. 17, Supplement A, 4-6 (2001).

DURABILITY OF THICK FILM CERAMIC GAS SENSORS

Adnan Merhaba and Sheikh Akbar
Department of Materials Science and Engineering
NSF Center for Industrial Sensors and Measurements (CISM)
291 Watts Hall, 2041 College Road
Columbus, OH 43201

Bin Feng and Golam Newaz
Mechanical Engineering Department
Wayne State University
2135 Engineering Bldg.
Detroit, MI 48202

Laura Riester and Peter Blau
Oakridge National Laboratory
Bldg. 4515, MS 6069
Oakridge, TN 37831-6069

ABSTRACT

Titanium dioxide in its anatase phase has been extensively studied for carbon monoxide sensing at The Center for Industrial Sensors and Measurements (CISM). Thick and thin film processing techniques such as screen-printing and spin coating have been used to fabricate these planar gas sensors. The issues related to mechanical integrity and adhesion between the film and the substrate (typically aluminum oxide) are generic to any planar gas sensors and they are addressed here. The critical balance between porosity (required for high sensitivity) and heat treatment (to achieve adequate bonding between the film and the substrate) can be a problem with most gas sensors. Bonding materials such as Tetra ethyl ortho-silicate (TEOS) have been tried to achieve good adhesion. Characterization techniques such as SEM, AFM and profilometry were carried out to learn about the morphology of the film. Various adhesion tests such as Thermal Wave Imaging, Nano-Indentation and Scratch tests have been carried out to get an indication (qualitative as well as quantitative) of adhesion. At times it is not always possible to correlate the results achieved from Thermal Wave Imaging tests to the adhesion of the film. On the other hand carrying out quantitative tests on porous films is not an easy task. The issues related to adhesion tests of poorly sintered films are addresses in this paper.

To the extent authorized under the laws of the United States of America, all copyright interests in this publication are the property of The American Ceramic Society. Any duplication, reproduction, or republication of this publication or any part thereof, without the express written consent of The American Ceramic Society or fee paid to the Copyright Clearance Center, is prohibited.

INTRODUCTION

Fabrication procedures used for gas sensors have seen tremendous growth over the last several years. Earlier gas sensors were fabricated using a tubular design. However batch production of such a sensor design was not easy. Advances in thick and thin film technology, arising primarily from research in the microelectronics industry, have resulted in transformation from the tubular design to planar design of gas sensors. In the planar design the sensing material is deposited onto a ceramic substrate (with electrodes screen-printed on it). Subsequently lead wires are attached to the film and the sensor is packaged to protect the material from degradation. Figure 1 shows a schematic diagram of a planar gas sensor. Adhesion of the sensing film to the substrate is the most critical issue with respect to such a system. This paper deals with adhesion related issues in planar design of ceramic gas sensors.

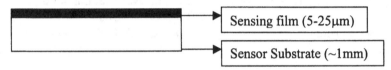

Figure 1 Schematic Diagram of a planar ceramic gas sensor.

Titanium dioxide in its anatase phase has been successfully developed as a carbon monoxide sensor at The Center for Industrial Sensors and Measurements (CISM).[1,2] The change in stoichiometry of titanium dioxide (as well as other semi-conducting oxides) as a function of the oxygen activity of their environment, particularly at elevated temperatures, is well known in literature. The change in stoichiometry can affect the electrical conductivity, σ, of the material (Kofstad, 1972):

$$\sigma = \sigma_0 \exp(E_A/kT)\, p(O_2)^{1/n} \qquad (1)$$

where k denotes the Boltzmann constant, T the temperature in degrees Kelvin, E_A is activation energy and the sign and value of n depends on the nature of the point defects arising when the oxygen is removed from the lattice. The term $p(O_2)$ is the partial pressure of oxygen in the environment. At high temperatures in a low oxygen environment, titania is an n-type semi-conducting oxide because of a substantial concentration of oxygen vacancies. When the metal oxide is heated at a high temperature, oxygen is adsorbed on the crystal surface with a negative charge. This leads to formation of dipole charge layers and associated depletion regions at the surface of each grain, which in turns results in back-to-back Schottky barriers. Thus the resistivity of the granular film is much higher than that of a bulk sample.[3] In the presence of a reducing gas such as CO, reactions at the grain surface leading to CO_2 liberate free electrons in the depletion layer, reducing the resistance at the interface and are the basis for the sensing mechanism. From the sensing mechanism described above, it is imperative to have a high surface to volume ratio of the sensing film or in other words it is important to have a high content of porosity in the film for good sensing. As we will see, this porosity requirement poses a big challenge with regard to quantification of adhesion.

EXPERIMENTAL PROCEDURE

The first step in fabricating the sensor is preparation of titanium dioxide paste. We start with very fine titanium dioxide (anatase) powder (Alfa Aesar 99.9% metals basis, 0.1-1.0 μm diameter). TiO_2 paste is prepared by adding organic vehicle (Heraeus Inc.) and solvent to dissolve the vehicle, to the powder. Dispersant (DISPERBYK 110 – BYK Chemie Inc.) is added to avoid flocculation of titania particles in the paste. Bonding material (TEOS, Alfa Aesar) in the range of 0 – 10% is added to the paste to improve adhesion and the paste is ultrasonically stirred for 60 minutes. The paste is screen-printed (Model MC810-C Thick Film Printer, C.W.Price Co. Inc.) or Spin Coated (Spin Coater Model P6700 series, Specialty Coating Systems) onto an aluminum oxide substrate (1.5cm x 1.5cm x 1mm, Laser Tech. Co.). The sensor is subsequently heat treated between 700 - 1000°C. The film is heat treated such that a high porosity content is achieved, which is essential for high sensitivity of the gas sensor.

Characterization of the sensor was done using SEM (XL – 30 FEG, Philips Co.) to learn about the morphology and the interface of the sample. Thickness and roughness of the film was measured using a Rodenstock non-contact laser profilometer. Surface quality and defects may be identified using Atomic Force Microscopy (AFM) with precision at the nanometer level. AFM is a non-destructive technique. AFM gains information based on atomic forces between a mechanical cantilever arm and the surface and can image a wide range of materials. In the contact mode, a cantilever with an atomically sharp tip is scanned over the sample surface. The surface deflections are recorded via a laser beam focused on the back of the cantilever. The beam reflects off the back of the cantilever onto segmented photodiode.

Residual stress calculations were carried out for the TiO_2-Al_2O_3 system. The thermal coefficient expansion of titania ($\alpha = 8.22 \times 10^{-6}$ K^{-1}) is slightly larger than that of alumina ($\alpha = 7.5 \times 10^{-6}$ K^{-1}).[4] Based on these numbers a residual tensile stress of approximately 32 MPa was calculated. Qualitative as well as quantitative adhesion tests were carried out to learn about the TiO_2 – Al_2O_3 interface. Non destructive tests such as Thermal Wave Imaging (TWI, EchoTherm® IR NDT, Thermal Wave Imaging Inc.) were employed to get a qualitative idea about the adhesion between the film and the substrate. During the TWI test, the sample is heated with a very high intensity Xe lamp and an Infrared camera monitors the cooling (thermal waves) of the surface. If the adhesion between the film and the substrate were good, the thermal diffusion would be faster to the substrate, via the interface, subsequently resulting in a lower temperature on the surface. On the other hand in the case of a poorly bonded film the thermal diffusion would be slower due to poor thermal conductivity of air, thereby resulting in a higher surface temperature. The TWI produces a graph of thermal amplitude (which is proportional to the surface temperature) as a function of time. A lower thermal amplitude indicates better bonding.

Other adhesion tests carried out were Scratch tests and Nano-indentation. These tests were carried out at the High Temperature Materials Laboratory (HTML) at the Oakridge National Laboratory. Scratch test (Model 502 motorized shear/scratch tester, Taber Industries) involves scratching the surface of the film with a diamond tip, while continuously increasing the load on the tip. At some critical load either the film would peel off from the substrate or the film would fracture. After careful examination of the failure mode, one can correlate the failure load with adhesion.[5,6] Nano-indentation (Nano Inedenter II, MTS Nano Instruments) involves penetrating the film with a diamond tip (Berkovich geometry) and subsequently measuring the penetration depth. A graph of load vs. displacement is plotted, which gives materials properties such as elastic modulus and hardness number. According to Sanchez, these properties are used during cross-sectional nano-indentation, which reveals the interfacial critical energy release rate, which is a direct measure of adhesion.[7]

RESULTS AND DISCUSSIONS

Morphology of the sensing film

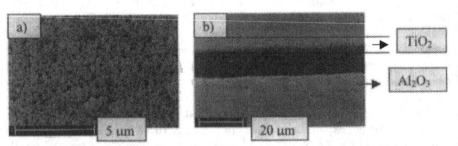

Figure 2 SEM Micrographs of Titanium dioxide a) surface and b) interface.

Figures 2(a) and 2(b) show the SEM micrographs of the surface and interface of the titanium dioxide sensor. As expected, the surface has a high content of porosity which is essential for high sensitivity of the sensor, however while mounting the sample in epoxy and subsequently grinding and polishing it,

procedure. Figures 5(a) and 5 (b) show surface profile information at the nanoscale over a small zone before and after 60 thermal cycles of the sensor. The x-axis of the graph represents the length of this line and the y-axis gives the surface undulations or surface asperity information at the nano scale. The surface undulations seem to have increased after 60 thermal cycles from an average of 0.8 nm to 2.5 nm.

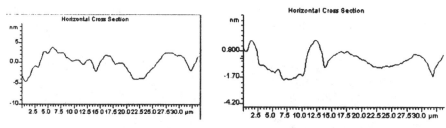

Figure 5 AFM surface profile information a) before and b) after 60 thermal cycles.

TWI Results

TWI tests were carried out on samples heat treated to temperatures ranging between 700 and 1000°C. Figure 6(a) shows the image captured by the IR camera during the test, 6(b) is the plot of thermal amplitude as a function of time for the 4 different samples and 6(c) is the 3-D thermal wave graph of the same 4 samples. The 3-D thermal graph takes into consideration the thermal diffusion across the breadth of the sample, thereby avoiding any ambiguity that may arise due to local defects. The results from the 3-D graphs are consistent with the ones in 6(b). As mentioned earlier lower thermal amplitude indicates better bonding, the results reveal that samples heat treated at 700°C shows better bonding than samples heat treated at 1000°C. However results obtained from TWI are not always unambiguous, because the thermal diffusion depends on the porosity within the film also. Thus quantitative adhesion test were tried to get a better idea about the system.

Nano-indentation test

As mentioned earlier, during nano-indentation a diamond tip indents the surface of the film to evaluate material properties such as elastic modulus and hardness. Unfortunately, in our case during indenting the tip almost sank into the film. It was similar to indenting through a pile of sand, primarily because of the powdery nature of the film. The test revealed that not only is the adhesion between the film and substrate critical, but also inter-particle bonding is crucial for long-term durability of the sensor.

delamination of the film from the substrate was observed, which already indicates that the adhesion for long-term durability is not good.

Figures 3(a) and 3(b) show the morphology of the film heat treated to 700 and 1000°C respectively. The sample heat treated at 1000°C shows higher porosity compared to the one heat-treated at 700°C, which seems counter-intuitive. However titanium dioxide undergoes phase transformation from anatase to rutile at a temperature range of 800-900°C (confirmed by X-ray diffraction results). From specific volumes of the two phases a 10-15% reduction is expected during the phase transformation which explains the observation of these micrographs.

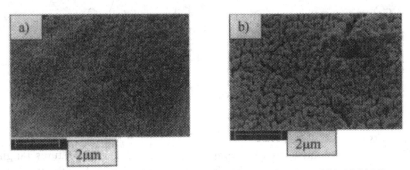

Figure 3 Morphology of TiO_2 film heat-treated to a) 700°C and b) 1000°C.

Figure 4 shows the results obtained from the profilometry tests. As expected, the thickness of the film increases linearly with the number of print cycles during screen-printing. However an average roughness of 2-3μm was recorded.

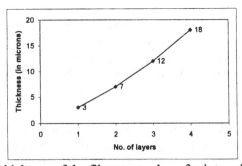

Figure 4 Graph of thickness of the film vs. number of print cycles.

This implies that if we have just one coating of the film the thickness and roughness are of comparable dimensions. Therefore to avoid any effect of variation in thickness along the length of the film on the adhesion tests it was recommended that atleast 3-4 print cycles be used during the fabrication

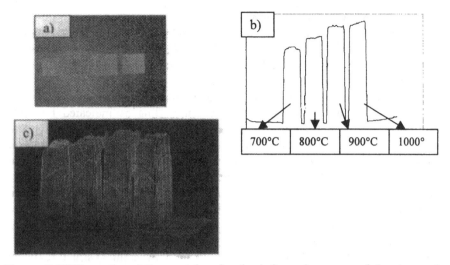

Figure 6 TWI results: a) Image taken by the infra red camera of the 4 samples heat treated to 700, 800, 900 and 1000°C (from left to right), b) Graph of thermal amplitude of the 4 corresponding samples as a function of time, c) 3-D thermal wave graph of the same 4 samples.

Scratch test

For the scratch test, samples with varying heat treatments (between 700°C and 1000°C), varying percentages of TEOS (0-10%) and varying thickness (5-25µm) were used. All the films failed at as low a load as 25 grams, which was the lowest load that could be applied on the scratch tester. Figure 7 shows the optical micrograph of two samples with a) 0% TEOS and b) 6% TEOS, both the samples were scratched with a load of 25 grams on the diamond tip. In both cases, the film failed at the lowest applicable load.

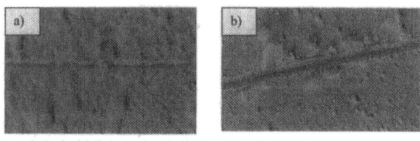

Figure 7 Optical Micrographs of samples after scratch test: a) 0% TEOS and b) 6% TEOS.

SUMMARY AND FUTURE WORK

The work reported in this paper deals with mechanical reliability and adhesion related issues of planar ceramic gas sensor. Various indications of the critical nature have been reported in the form of SEM micrographs of the interface and Scratch test results. Thickness and roughness measurements indicate that atleast 3-4 print cycles should be used during the fabrication of the sensor. Surface quality seems to have decreased after thermal cycling as indicated by AFM results. Thermal Wave Imaging tests have been carried out to qualitatively get an idea of adhesion between the film and the substrate. Due to ambiguity in interpreting the results, quantitative tests such as nano indentation were tried out. However poor inter particle bonding within the film restricted the application of such tests on TiO_2-Al_2O_3 system. Various parameters such as heat treatment, percentage of TEOS and different thickness of the film have been tried to improve adhesion. Currently we are working on a novel technique to fabricate the CO sensor using sputtering and photo electro chemical etching of titanium dioxide. Photo etching of titanium dioxide produces a nano-honeycombed structure, which would be an ideal platform for gas sensing application.[8,9] Future work also includes use of reactive bonding, in the form of CuO in lieu of TEOS. It has been reported that CuO increases selectivity to CO.[1] CuO also forms a spinel with Al_2O_3 at 800°C, which we believe will improve the adhesion between TiO_2 and Al_2O_3. Theoretical techniques based on fracture mechanics formulation are being pursued on TiO_2-Al_2O_3 system to estimate the interfacial fracture toughness of the film.

Acknowledgments

Research at the Ohio State University and Wayne State University was supported through NSF Grants (EEC-9872531 and EEC-9523358). Research at the Oakridge National Laboratory was sponsored by the Assistant Secretary for Energy Efficiency and Renewable Energy, Office of Transportation Technologies, as part of the High Temperature Materials Laboratory User Program, Oak Ridge National Laboratory, Managed by UT-Battelle, LLC, for the U.S. Department of Energy under contract number DE-AC05-00OR22725.

REFERENCES

[1] Dutta, P., et al., "Interaction of Carbon Monoxide with Anatase Surfaces at High Temperatures: Optimization of a Carbon Monoxide Sensor," *J. Phys. Chem. B*, pp. 4412-4422, 1999.

[2] Savage, N., et al., "Titanium Dioxide Based High Temperature Carbon Monoxide Selective Sensor," *Sensors and Actuators B*, 72, pp. 239-248. 2001.

[3] Akbar S. A., Younkman L. B., "Sensing Meachnism of a Carbon Monoxide Sensor Based on Anatase Titania," *J. Electrochem. Soc.* 144: 1750-53, 1997.

[4]Shackelford, J., et al., *CRC Materials Science and Engineering Handbook*, Boca Raton: CRC Press, 1994.

[5]Bull, S.J., *Materials at High Temperatures* Volume 13, No. 4, pp. 169-174. 1995.

[6]Thouless, M.D., *Engineering Fracture Mechanics*, 61, pp. 75-81, 1998.

[7]Sanchez, J., et al., *Acta matter* Vol. 47, No. 17, pp. 4405-4413, 1999.

[8]Sugiura, T., et al., "Microstructural Observation of Photoelectrochemically Tailored Nano-Honeycomb TiO_2," *Electrochemistry (Japanese)*, 67C, No. 12, 1, 1999.

[9]Sugiura, T., "Designing a TiO_2 Nano-Honeycomb Structure Using Photoelectrochemical Etching," *Electrochemical and Solid State Letters*, 1(4), 175-177, 1998.

PREPARATION AND CHARACTERIZATION OF INDIUM-DOPED CALCIUM ZIRCONATE FOR THE ELECTROLYTE IN HYDROGEN SENSORS FOR USE IN MOLTEN ALUMINUM

A.H. Setiawan
Materials Science Program, University of Indonesia
Jl. Salemba 4, Jakarta 10430, Indonesia

J.W. Fergus
Materials Research and Education Center, Auburn University
201 Ross Hall, AL 36849-5341, USA

ABSTRACT

$CaZrO_3$ doped with indium may be used as the solid electrolyte in hydrogen sensors. Hydrogen in molten aluminum during casting can lead to porosity, so hydrogen is removed by degassing. Measurement of the amount of hydrogen can be used to optimize the degassing process and thus improve the quality of the resulting aluminum casting. In this work, two preparation methods (powder and liquid-mix processing) were utilized to synthesize $CaZr_{1-x}In_xO_{3-x/2}$. The effect of dopant concentration on the structure was determined using x-ray powder diffraction and the conductivity in a pure argon atmosphere was determined using impedance spectroscopy.

INTRODUCTION

Hydrogen in Molten Aluminum

Hydrogen can be dissolved in molten aluminum when moisture in the atmosphere reacts with the melt to form aluminum oxide and hydrogen, which then diffuses into the melt[1]. The solubility of hydrogen in molten aluminum is higher than that in solid aluminum, so hydrogen can be evolved during solidification and lead to porosity. Therefore, the amount of dissolved hydrogen must be minimized before casting. Techniques for removal of hydrogen from aluminum alloy melts include natural outgassing, vacuum degassing and bubble degassing. The basic mechanism for these techniques is to establish above the molten aluminum a hydrogen partial pressure which is lower than that in equilibrium with the molten aluminum[2]. Measurement of the actual hydrogen

---

To the extent authorized under the laws of the United States of America, all copyright interests in this publication are the property of The American Ceramic Society. Any duplication, reproduction, or republication of this publication or any part thereof, without the express written consent of The American Ceramic Society or fee paid to the Copyright Clearance Center, is prohibited.

content in the melt using *in situ* hydrogen sensors can be used to optimize the degassing process[3].

Electrochemical Sensors

One approach to measuring hydrogen in molten aluminum is the use of sensors based on solid electrolytes (most commonly proton-conducting electrolytes). Some hydrogen sensors have been designed for low temperatures using electrolytes such as antimonic acid and hydrogen uranyl phosphate (UHP)[4], but these materials do not have the high-temperature stability to be used in contact with molten aluminum. One approach to developing new proton conductors has been to modify sodium-ion-conducting materials, such as zirconium phosphate (NASICON), NASICON-based materials (Zirpsio) and β/β"-alumina, so that they conduct protons. Some of these electrolytes do not have sufficient temperature stability and the presence of sodium in the aluminum melt may cause interference[5-10]. Oxide-ion conductors have also been used, but the sensor output depends on a mixed potential between hydrogen, oxygen and water vapor, which may cause difficulties in interpretation and also may lead to interference from the atmosphere[11].

Over the past several years, a new class of proton-conducting oxides has been developed. The best of these are based on strontium and barium cerate[12]. Although these proton-conducting oxides have been used in hydrogen sensors, they suffer from instability[13]. Zhuijkov[14,15] has recently reviewed potential proton conducting materials for high temperature applications and identified barium cerate and calcium zirconate as materials which could be used above 500°C. Iwahara's research group, which has worked on a wide variety of proton-conducting oxides, has developed a hydrogen sensor based on calcium zirconate doped with indium[16]. The $CaZrO_3$ has been doped with indium, because the ionic radius of indium is close to that of zirconium, and leads to high conductivity[13]. Powder processing has commonly been used for preparation of the solid electrolyte materials[13,16,17]. However, ceramic synthesis processes based on organic precursors can reduce processing times and temperatures[18,19] and have been used for synthesis of $CaZrO_3$[20-22] based materials.

In this work, the calcium zirconate doped with indium has been prepared by either powder processing or liquid-mix processing and has been analyzed using x-ray diffraction and impedance spectroscopy.

EXPERIMENTAL
Materials

Calcium carbonate ($CaCO_3$, Alpha Aesar, 99.5% metal basis, 5 μm powder), indium (III) oxide (In_2O_3, Alpha Aesar, 99.99% metal basis, -325 mesh powder), zirconium oxide (ZrO_2, Alpha Aesar, 99+% metal basis, -325 mesh) and zirconyl chloride octahydrate ($ZrOCl_2 \cdot 8H_2O$, 99.5%) were used as the starting materials. Oxalic acid dihydrate ($C_2H_2O_4 \cdot 2H_2O$, Sigma Chemical) and polyethylene glycol 200 and 1450 (Sigma Chemical) were used as reactants and surface area agents.

Materials Preparation

$CaZr_{1-x}In_xO_{3-x/2}$ samples were prepared using either powder processing or liquid-mix processing. For the powder processing, calcium carbonate, indium oxide and zirconium oxide were mixed in the stoichiometric ratio for $x=0.1$ with ethanol and then ball-milled for about 24 hours to produce a slurry. The slurry was then evaporated and dried in an oven at 100°C for 12 hours. The dried sample was calcined at 1000°C for 12 hours and then ground with a mortar and pestle.

The liquid-mix process was used to prepare oxides with x values of 0, 0.05, 0.10 and 0.20. Indium oxide was dissolved in a solution of hydrochloric acid and distilled water (1:1 ratio) and heated to about 90°C for 15 minutes. After dissolution of the indium, calcium carbonate was added to the solution. Zirconyl chloride octahydrate and polyethylene glycol 200 were separately dissolved in distilled water and added to the solution. The resulting solution was heated to 90°C and stirred for 30 minutes. Crystalline oxalic acid dihydrate dissolved in distilled water and polyethylene glycol 1450 dissolved in ethanol were mixed and the pH of this solution was adjusted to between 8 and 10 using a solution of ammonia and water (1:1 ratio). The two solutions were combined, and, after readjusting the pH to between 8 and 10, the solution was stirred for another 30 minutes at 90°C to precipitate the cations. The solution was aged for 4 hours and the resulting precipitate was washed with distilled water and filtered. After drying at 100°C for 12 hours the powder was calcined at 1000°C for 3 hours. The calcined sample was then ground by mortar and pestle.

For both methods, the powders were pressed at 240 MPa into cylindrical bars (5 cm length and 1 cm diameter) and then sintered in air at temperatures from 873°C to 1400°C. Finally, the sintered bars were sliced into 2-mm thick discs with a diamond saw and were examined using x-ray diffraction and impedance spectroscopy.

Chemical analysis was not performed on the samples after synthesis, so the compositions given below are nominal compositions based on the amounts of starting materials.

Measurement Techniques

X-ray diffraction. The phase contents in the samples were analyzed using x-ray diffraction. The measurements were carried out using a Rigaku diffractometer with CuK_α radiation for 2θ from 20° to 70°.

Impedance spectroscopy. For electrical measurements, both flat faces of the discs were polished and painted with a thin layer of carbon ink to form the electrodes. The samples were placed between nickel electrodes and placed in a spring-loaded sample holder. In all measurements, the electrode areas and the thicknesses of samples were the same. The electrical resistance measurements were carried out under argon atmosphere in the frequency range of 10 Hz to 32 MHz using impedance spectroscopy (1260 Solarton Impedance/Gain-Phase Analyzer) at temperatures from 500°C to 900°C.

RESULTS AND DISCUSSION
Phase Transformations During Processing

The XRD results for the samples are given in Figure 1 and Figure 2 and confirm that $CaZrO_3$ was formed by both methods. For the sample prepared by powder processing (Figure 1) a small peak at 33.42° was observed after sintering at 1400°C for only 10 hours. The peak was indexed to $CaIn_2O_4$ and was not present after sintering for 20 hours at the same temperature indicating that indium oxide dissolved into the calcium zirconate structure.

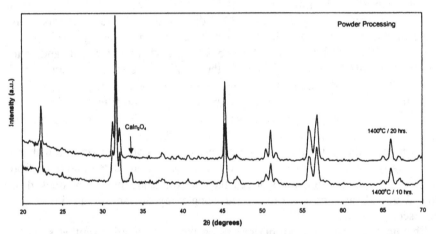

Figure 1. XRD Spectra for Sample Prepared by Powder Processing

The XRD results for samples prepared by the liquid-mix process (Figure 2) show that after heating at temperatures of 100°C for 12 hours to 873°C for 3 hours, $CaZrO_3$ had not yet formed. However after heating to 940°C to 1000°C for 3 hours, $CaIn_2O_4$ and $CaZrO_3$ formed. The $CaIn_2O_4$ peak then disappeared after 12 hours at 1000°C indicating that indium oxide dissolved into the $CaZrO_3$ structure. Thus, the liquid-mix process can be used to synthesize $CaZr_{1-x}In_xO_{3-x/2}$

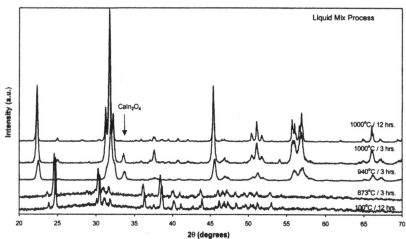

Figure 2. XRD Spectra for Sample Prepared by Liquid Mix Processing

with a lower sintering temperature and/or a shorter sintering time as compared to powder processing.

Conductivity Measurements

The equivalent circuit and the measured impedance spectra at 700°C for the sample with $x=0.10$ prepared by the liquid-mix process are shown in Figure 3. The circles represent the experimental data, and the line represents values calculated from a curve-fitting program using the equivalent circuit. The impedance spectra of other samples were of similar shape to that shown in Figure 2. In the following analysis, the high frequency arc is assumed to represent the impedance of the grain bulk and at the lower frequency arc is assumed to represent the impedance of the grain boundary. In the equivalent circuit, CPE represents a constant phase element.

The conductivity of grain boundary and bulk. The temperature dependences of the grain boundary and bulk conductivities of the sample of $CaZr_{0.9}In_{0.1}O_{2.95}$ prepared by powder processing, as an example, are shown in Figure 4. Both conductivities follow an Arrhenius relationship from 500°C to 800°C. The activation energy of the bulk conductivity is higher than that of the grain boundary, so that bulk grain conduction dominates at higher temperatures.

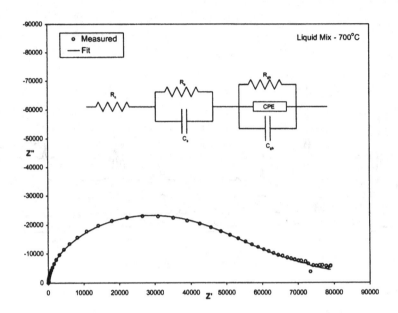

Figure 3. Typical Impedance Spectra and Equivalent Circuit.

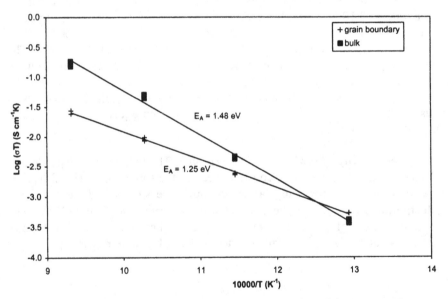

Figure 4: Grain Boundary and Bulk Conductivity of $CaZr_{0.9}In_{0.1}O_{2.95}$ Prepared by Powder Processing.

Effect of processing method on conductivity. The total conductivities from total resistivities for samples prepared by both methods are shown in Figure 5. The magnitude and temperature dependence of the total conductivies for samples prepared by both methods are similar.

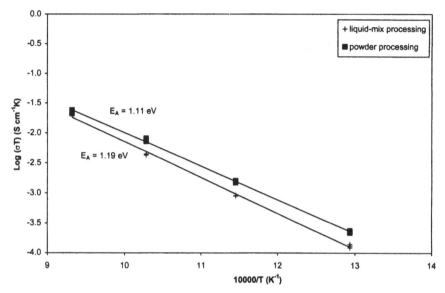

Figure 5: Effect of Preparation Method on Total Conductivity of $CaZr_{0.9}In_{0.1}O_{2.95}$.

Effect of composition on conductivity: The conductivities calculated from the total resistances at different compositions (x = 0, 0.05, 0.10, 0.20) are shown in Figures 6 and 7. The increase in conductivity with increasing indium content is presumably due to an increase in the oxide-ion vacancy content according to Reaction 1.

$$2V_{Zr}^{////} + 4V_O^{\bullet\bullet} + In_2O_3 = 2In_{Zr}^{/} + V_O^{\bullet\bullet} + 3O_O^x \tag{1}$$

The line shown is fit to the data for the indium-containing samples and shows that the marginal increase in conductivity decreases with increasing indium content.

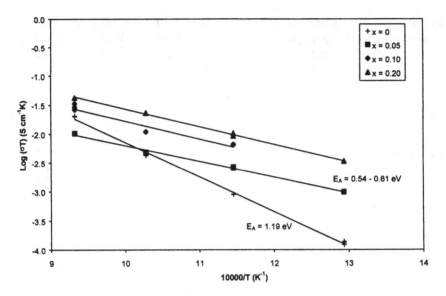

Figure 6: Effect of Composition on Total Conductivity and Activation Energy of $CaZr_{1-x}In_{0.1}O_{3-x/2}$.

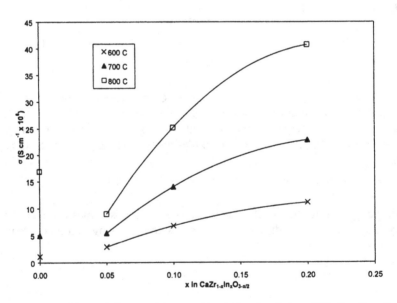

Figure 7: Effect of Composition on Total Conductivity of $CaZr_{1-x}In_{0.1}O_{3-x/2}$ at 600°C to 800°C.

CONCLUSIONS
1. Preparation of indium-doped calcium zirconate by the liquid-mix process can be used to synthesize $CaZr_{1-x}In_xO_{3-x/2}$ with reduced sintering time and temperature as compared to its preparation by powder processing.
2. The preparation method does not significantly affect the total conductivity.
3. The conductivity of $CaZr_{1-x}In_xO_{3-x/2}$ for $x=0$ to 0.20 increases as the amount of In_2O_3 increases, presumably due to an increase in the concentration of oxide-ion vacancies.

ACKNOWLEDGEMENT
This work was financially supported by NASA through the Solidification Design and Control Consortium.

REFERENCES
[1] M.M. Makhlouf, L. Wang and D. Apelian, *Measurement and Removal of Hydrogen in Aluminum Alloys,* American Foundrymen's Society, Des Plaines, IL, 1-6 (1998).

[2] G.K. Sigworth and T.A. Engh, "Chemical and Kinetic Factors Related to Hydrogen Removal from Aluminum," *Metall. Trans B,* 13B 447-460 (1982).

[3] X.G. Chen, F.J. Klinkenberg, S. Engler, L. Heusler and W. Schneider, "Comparing Hydrogen Testing Methods for Wrought Aluminum," *JOM,* 46 [8] 34-38 (1994).

[4] R.V. Kumar and D.J. Fray, "Development of Solid-State Hydrogen Sensors," *Sensors and Actuators,* 15 [2] 185-191 (1988).

[5] L.B. Kriksunov and D.D. Macdonald, "Amperometric Hydrogen Sensor for High-Temperature Water," *Sensors and Actuators B,* 32 157-160 (1996).

[6] L.D. Angelis, A. Maimone, L. Modica, G. Alberti and R. Palombari, "A New Hydrogen Sensor with Pellicular Zr Phosphate as Proton Conductor," *Sensors and Actuators B,* 1 121-124 (1990).

[7] S.F. Chehab, J.D. Canaday, A.K. Kuriakose, T.A. Wheat and A. Ahmad, "A Hydrogen Sensor Based on Bonded Hydronium NASICON," *Solid State Ionics,* 45 [3-4] 299-310 (1991).

[8] J. Gulens, T.H. Longhurst, A.K. Kuriakose and J.D. Canaday, "Hydrogen Electrolysis Using A Nasicon Solid Protonic Conductor," *Solid State Ionics,* 28-30 622-626 (1988).

[9] J.D. Canaday, A.K. Kuriakose, A. Ahmad and T.A. Wheat, "Device Applications of Hydrogen-Conducting Solid Electrolytes," *J. Can. Ceram. Soc.,* 55 34-37 (1986).

[10] M. Dekker, I.'t Zand, J. Schram and J. Schoonman, "NH_4Y and HY Zeolites as Electrolytes in Hydrogen Sensors," *Solid State Ionics,* 35 [1-2] 157-164 (1989).

[11] G. Lu, N. Miura and N. Yamazoe, "High-Temperature Hydrogen Sensor Based on Stabilized Zirconia and a Metal Oxide Electrode," *Sensors and Actuators B*, 35-36 130-135 (1966).

[12] J. Luyten, F. DeShutter, J. Schram and J. Schoonman, "Chemical and Electrical Properties of Yb-Doped Strontium Cerates in Coal Combustion Atmospheres," *Solid State Ionics*, 46 [1-2] 117-120 (1991).

[13] T. Yajima, H. Kazeoka, T. Yogo and H. Iwahara, "Proton Conduction in Sintered Oxides Based on $CaZrO_3$," *Solid State Ionics*, 47 271-275 (1991).

[14] S. Zhuiykov, "Development High-temperature Hydrogen Sensor Based on Pyrochlore Type of Proton-conductive Solid Electrolyte," *Ceram. Eng. Sci. Proc.*, 17 [3] 179-186 (1996).

[15] S. Zhuiykov, "Hydrogen Sensor Based on A New Type of Proton Conductive Ceramic," *Int. J. Hydrogen Energy*, 21 [9] 749-759 (1996).

[16] T. Yajima, H. Iwahara, K. Koide and K. Yamamoto, "$CaZrO_3$-type Hydrogen and Steam Sensors: Trial Fabrication and Their Characteristics," *Sensors and Actuators B*, 5 145-147 (1991).

[17] N. Kurita, N. Fukatsu, K. Ito and T. Ohashi, "Protonic Conduction Domain of Indium-doped Calcium Zirconate," *J. Electrochem. Soc.*, 142 [5] 1552-1559 (1995).

[18] N.G. Error and H.U. Anderson, "Polymeric Precursor Synthesis of Ceramic Materials," *Mat. Res. Soc. Symp.Proc.*, 73 571-577 (1986).

[19] H.U. Anderson, M.M. Nasrallah and C.C. Chen, "Method of Coating A Substrate with a Metal Oxide Film from an Aqueous Solution Comprising A Metal Cation and a Polymerizable Organic Solvent," U.S. Pat. No. 5 494 700, Feb. 27, 1996.

[20] M. Rajendran and M.S. Rao, "Strontium and Calcium Zirconyl Citrates as Precursors for the Low-Temperature Synthesis of $SrZrO_3$ and $CaZrO_3$ Fine Powder," *J. Mater. Res.*, 12 [10] 2665-2672 (1997).

[21] Y. Wei, L. Guangqiang and S. Zhitong, "Coprecipitating Synthesis and Impedance Study of $CaZr_{1-x}In_xO_{3-\delta}$," *J. Mat. Sci. Letters*, 17 241-43 (1998).

[22] I.E. Gonenli and A.C. Tas, "Chemical Synthesis of Pure and Gd-Doped $CaZrO_3$ Powders," *J. Eur. Ceram. Soc.*, 19 [13-14] 2563-2567 (1999).

ANTIMONY SENSOR FOR MOLTEN LEAD USING K-β-Al$_2$O$_3$ SOLID ELECTROLYTE

Rajnish Kurchania[a] and Girish. M. Kale[b]
School of Process Environmental and Materials Engineering
Department of Mining and Mineral Engineering
University of Leeds, Leeds LS2 9JT, UK

ABSTRACT

In any metal refining operation, on-line monitoring of dissolved impurities in molten metal is very important for minimising the production of excessive dross and improving the energy efficiency of the process. In this paper we report the synthesis and charaterisation of materials used in designing the sensor as well as the fabrication and testing of the antimony sensor in molten lead at 923 K. The results of the physical and electrical characterisation of the solid electrolyte material K-β-Al$_2$O$_3$ have been presented.

INTRODUCTION

Antimony is present as a dissolved impurity element in lead produced by pyrometallurgical route. Antimony is an undesirable alloying element in lead and therefore needs to be detected and removed during the refining process. An antimony sensor capable of operating in the temperature range 873-973 K will be very useful for continuously monitoring the lead refining process. In view of this technological need, the present investigation has been undertaken for the development of solid-state antimony sensor for molten lead. Kurchania and Kale have published a detailed review on the electrochemical sensors for molten metal[1]. This research work is part of a major ongoing research program in our laboratory on the design and development of solid-state electrochemical sensors for molten metal. Oxygen sensor for molten silver and tin has been successfully developed in our laboratory[2,3]. In the present investigation attempts have been made to synthesise and characterise a novel potassiun-beta-alumina+potassium

---

[a] Presently at the Materials Research Centre, Department of Engineering & Applied Science, University of Bath, Bath BA2 7AY, UK
[b] Member American Ceramic Society

To the extent authorized under the laws of the United States of America, all copyright interests in this publication are the property of The American Ceramic Society. Any duplication, reproduction, or republication of this publication or any part thereof, without the express written consent of The American Ceramic Society or fee paid to the Copyright Clearance Center, is prohibited.

antimony oxide, (K-β-Al$_2$O$_3$+KSbO$_3$) based antimony ion conductor for the fabrication of antimony sensor for molten lead. The refractory nature of K-β-Al$_2$O$_3$ based solid electrolyte ceramic material makes them capable of accurately and instantaneously monitoring the dissolved impurities in molten metal in hostile and corrosive environments. Kale et. al.[4] have successfully developed a sodium – beta – alumina + sodium antimony oxide, (Na-β-Al$_2$O$_3$+ NaSbO$_3$) based antimony ion conducting solid electrolyte material for sensing antimony in molten tin, copper and zinc. Yttria stabilised zirconia (YSZ) based antimony sensor for molten zinc was reported by Fergus et. al.[5]. However, these antimony sensor[5] are more suitable for sensing higher antimony contents and therefore it is also required to address the issue of stability of auxiliary phase in the dilute region of zinc-antimony alloy (Sb ≤ 1 wt %) because the technological importance of antimony sensor during metal refining is mainly in the dilute region. Therefore, in the present investigation attempts have been made to develop a long-life and robust antimony sensor which is capable of detecting antimony in the molten lead in the dilute region (Sb ≤ 1 wt %) as accurately as possible.

EXPERIMENTAL PROCEDURE
(i) Synthesis of solid electrolyte, reference electrode and auxiliary phase material
Potassium beta alumina: Three compositions of i.e. K$_2$CO$_3$: Al$_2$O$_3$ (1 : 5.25, 6.69 and 9) have been synthesised by solid solution ceramic processing route. Prior to mixing the potassium carbonate (K$_2$CO$_3$) was dried at 473 K for 12 hours and then the mixture of K$_2$CO$_3$ and Al$_2$O$_3$ was ground under acetone for 1 hour. Then the slurry was dried at 343 K for 4 hours. The powder was pre-calcined at 1073 K for 6 hours and at 1373 K for 10 hours. After wet mixing and drying the final calcination was done at 1573 K for 10 hours.

Potassium beta ferrite: Prior to mixing the potassium carbonate (K$_2$CO$_3$) was dried at 473 K for 12 hours and then the mixture of K$_2$CO$_3$ and Fe$_2$O$_3$ was ground under acetone for 1 hour. Then the slurry was dried at 343 K for 4 hours. The powder was pre-calcined at 1173 K for 1 hour and at 1473 K for 5 hours. After wet mixing and drying the final calcination was done at 1473 K for 5 hours.

Potassium antimony oxide: Prior to mixing the potassium carbonate (K$_2$CO$_3$) was dried at 473 K for 12 hours and then the mixture of K$_2$CO$_3$ and Sb$_2$O$_3$ was ground under acetone for 1 hour. Then the slurry was dried at 343 K for 4 hours. The powder was calcined at 1273 K for 20 hours, using the slow heating schedule i.e. increasing temperature by 373 K and hold time was about 15 hours at each temperature.

(ii) % weight loss, density and conductivity measurements

Pellets of all three compositions of K-β-Al$_2$O$_3$ having dimensions of 10 mm diameter and 2 mm thickness have been iso-statistically pressed at 200 MPa and sintered at 1873 K for 30, 60 and 120 min. The weight of K-β-Al$_2$O$_3$ pellets before and after sintering has been measured and then the percentage weight loss in K-β-Al$_2$O$_3$ pellets has been calculated. One end-closed tubes (having dimensions 26 mm long × 4 mm inner diameter × 6 mm outer diameter) of composite solid electrolyte (potassium-beta-alumina + potassium antimony oxide) have also been prepared and used as solid electrolyte tubes for the fabrication of antimony sensor.

Density of the sintered K-β-Al$_2$O$_3$ pellets has been measured at ambient temperature using helium gas pycnometer (Micromeritrics, AccuPyc 1330).

A.C. conductivity of the K-β-Al$_2$O$_3$ pellets have been measured as a function of increasing temperature from 473 K to 873 K, using impedance analyser (SI 1260, Schlumberger, Germany) in the frequency range of 1 Hz to 10 MHz. The sintered pellets were coated with silver paint (Agar Scientific, UK) to make the electrical contacts.

(iii) Electromotive force (EMF) measurements and design of antimony sensor

The antimony sensor was tested in molten lead in inert atmosphere at 923 K. Electromotive force (EMF) as a function of antimony level has been recorded using Keithley 6517A (Keithley Instruments, Inc., USA) high impedance (z ~ 200 TΩ) multi-channel digital electrometer driven by the Test Point Software (Capital Equipment Corporation, USA).

The solid-state antimony sensor used for measuring antimony in Pb-Sb system can be schematically shown as:

$$(+)\text{Fe-Cr, K-}\beta\text{-Fe}_2\text{O}_3+\text{Fe}_2\text{O}_3 // \text{K-}\beta\text{-Al}_2\text{O}_3+\text{KSbO}_3 // \underline{\text{Pb}}_{\text{Sb}}, \text{Fe-Cr}(-) \qquad (I)$$

The design details of the antimony sensor for molten lead are shown in Figure 1, which mainly consists of:

a) Sensor: This is made up of composite solid electrolyte tube (potassium-beta-alumina + potassium antimony oxide) filled with a mixture of potassium-beta-ferrite + iron oxide powder as reference electrode material connected electrically by Fe-Cr wire. This is fixed to a recrystallised alumina tube using ceramic cement.

b) Counter electrode: Molten lead-antimony (Pb-Sb) alloy in which the antimony level ranging from 0.02 to 1.00 wt %.

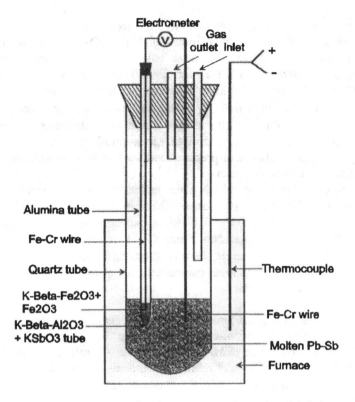

Figure 1. Design of antimony sensor for molten lead.

c) Contact lead: Fe-Cr wire held inside a tightly fitting high density alumina tube to provide electrical contact to Pb-Sb molten alloy.

d) Keithley 6517A Electrometer: To record the electromotive force (EMF) of the antimony sensor as a function of antimony addition in pure lead (Pb). The EMF values were recorded at 923 K, after about 30 min of each antimony addition in lead, so that the equilibrium can be attained and thereby stable EMF values can be recorded. In all these measurements it was observed that the EMF remains stable for about 15 min. The stable EMF values are considered for plotting the EMF as a function of natural logarithm of mole fraction of antimony in molten lead-antimony alloy. The Pb-Sb alloy samples were used for the accurate determination of the antimony contents in the alloy by Induction Coupled Plasma (ICP) analysis. A K-type thermocouple was placed adjacent to the alumina crucible holding the molten metal in order to record the exact temperature during the EMF measurements.

RESULTS

Three compositions of potassium-beta-alumina were synthesised by solid solution route (K_2CO_3 : Al_2O_3 1:5.25, 1:6.69 and 1:9) for their use as solid electrolyte material. XRD analysis confirms the formation of potassium-beta-alumina[6]. Potassium-beta-ferrite ($K_{1.33}Fe_{11}O_{17}$) has also been synthesised by solid solution route to use as reference electrode material. XRD analysis confirms the formation of potassium beta ferrite and iron oxide[7, 8].

Pellets of the three compositions of K-β-Al_2O_3 having dimensions of 10 mm diameter and 2 mm thickness, have been pressed at 200 MPa using iso-static press and sintered at 1873 K for 30, 60 and 120 min. In order to select the best composition for the fabrication of solid electrolyte tubes and also to optimise the solid electrolyte tube processing conditions, following investigations have been carried out (i) percentage (%) weight loss as a function of sintering time, (ii) density as a function of sintering time and (iii) ac conductivity as a function of temperature as shown in Table I, II and III and plotted in Figure 2, Figure 3, and Figure 4 respectively.

Table I. Percentage weight loss and sintering time for K-β-Al_2O_3 sintered at 1873 K

Sintering Time (min).	1 : 5.25 (K_2CO_3 : Al_2O_3)	1 : 6.69 (K_2CO_3 : Al_2O_3)	1 : 9 (K_2CO_3 : Al_2O_3)
30	6.80	6.10	1.08
60	10.20	8.50	1.61
120	11.37	10.28	2.61

Table II. Density and sintering time for K-β-Al_2O_3 pellets sintered at 1873 K

Sintering Time (min).	Density (gm cm^{-3}) 1 : 5.25 (K_2CO_3 : Al_2O_3)	Density (gm cm^{-3}) 1 : 6.69 (K_2CO_3 : Al_2O_3)	Density (gm cm^{-3}) 1 : 9 (K_2CO_3 : Al_2O_3)
30	3.3808	3.3933	3.4010
60	3.3595	3.3673	3.3709
120	3.3380	3.3400	3.3413

Table III. Conductivity as a function of reciprocal of temperature

Sintering Temperature $10^3/T$ (K^{-1})	log σ (ohm m)$^{-1}$ 1 : 5.25 (K_2CO_3 : Al_2O_3)	log σ (ohm m)$^{-1}$ 1 : 6.69 (K_2CO_3 : Al_2O_3)	log σ (ohm m)$^{-1}$ 1 : 9 (K_2CO_3 : Al_2O_3)
2.114	-4.477	-4.725	-5.176
1.745	-3.412	-3.741	-3.875
1.485	-2.614	-2.954	-3.000
1.293	-2.301	-2.602	-2.698
1.145	-2.000	-2.371	-2.498

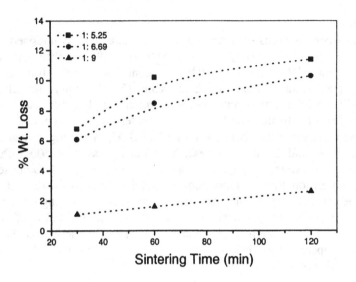

Figure 2. Percentage weight loss as a function of sintering time for K-β-Al_2O_3 sintered at 1873 K

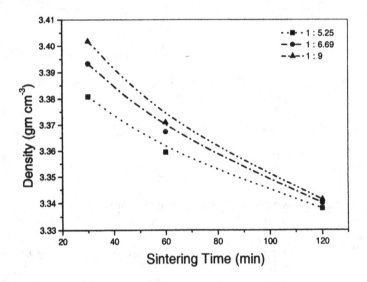

Figure 3. Effect of sintering time on densification of K-β-Al_2O_3 pellets sintered at 1873 K

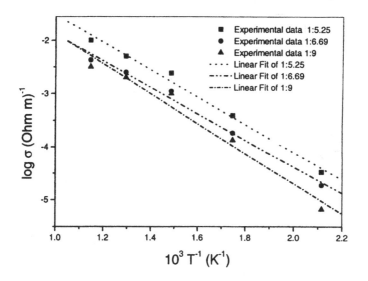

Figure 4. Conductivity as a function of reciprocal of temperature for K-β-Al_2O_3 pellets sintered at 1873 K

In order to investigate the effect of antimony (Sb) addition on the EMF of antimony sensor in molten Pb-Sb alloy. The EMF of antimony sensor was measured at 923 K in Pb-Sb alloy with increasing antimony (Sb) level from 0.02 to 1.00 wt % respectively. The measured EMF of the antimony sensor is then plotted as a function of natural logarithm of mole fraction of Sb in Pb (ln X_{Sb}) at 923 K as shown in Figure 5, where the mole fraction of Sb in Pb is defined as follows:

X_{Sb}=(wt.of Sb/At.wt.of Sb)/[(wt.of Sb/At.wt. of Sb) + (wt.of Pb/At.wt. of Pb)] (1)

Table IV. EMF as a function of lnX_{Sb} for antimony sensor tested at 923 K in Pb-Sb alloy

Wt % Sb	lnX_{Sb}	EMF (V)
0.05	-7.07	0.600
0.3	-5.28	0.430
0.5	-4.77	0.350
1.0	-4.08	0.250

Chemical Sensors for Hostile Environments

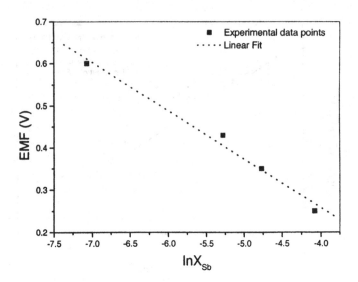

Figure 5. Variation in measured EMF as a function of ln X_{Sb} at 923 K for molten Pb-Sb alloy

A continuous decrease in the EMF from 0.600 V for 0.05 wt % Sb to 0.250 V for 1.00 wt % Sb was observed in molten lead-antimony alloy at 923 K as shown in Figure 5, this indicates that the sensor is following the Nerstian behaviour. It was observed that the time taken to attain chemical equilibrium was about 30 min and EMF remains stable for about 15 min. The antimony sensor used in the present investigations has been fabricated using the composite solid electrolyte tube synthesised in our laboratory.

The chemical analysis (ICP) of the Pb-Sb alloy samples was performed at Britannia Refined Metals and was found to be in good agreement with the antimony additions done in the laboratory.

DISCUSSION

For the fabrication of antimony sensor, K-β-Al_2O_3 based antimony ion conducting solid electrolyte material has been successfully synthesized and used for the fabrication of antimony sensor. In the present investigation three compositions of K-β-Al_2O_3 and $KSbO_3$ has been synthesised via solid solution route. Powder of 1:9 K-β-Al_2O_3 and $KSbO_3$ have been mixed heterogeneously in 9:1 ratio and subsequently their tubes have been pressed and used to fabricate the novel composite solid electrolyte for the antimony sensor. Potassium-beta-ferrite

(K-β-Fe$_2$O$_3$) has also been synthesised via solid solution route and a mixture of K-β-Fe$_2$O$_3$ and Fe$_2$O$_3$ was used as reference electrode material for antimony sensor. XRD analysis confirms the formation of the desired phases.

In order to select the best composition of K-β-Al$_2$O$_3$ for the fabrication of novel composite solid electrolyte tubes, the percentage (%) weight loss and density as a function of sintering time (30, 60 and 120 min at 1873 K) and conductivity as a function of temperature has been measured. It can be seen from Figure 2, that minimum % weight loss was observed in 1:9 K-β-Al$_2$O$_3$ composition sintered at 1873 K for 30 min. Figure 3, shows that maximum density was observed in 1:9 K-β-Al$_2$O$_3$ composition sintered at 1873 K for 30 min. The possible reason for such trends may be the higher volatility of potassium oxide in 1:5 and 1:6.69 K-β-Al$_2$O$_3$ compositions in comparison to that of 1:9 K-β-Al$_2$O$_3$ compositions, because the amount of K$_2$CO$_3$ present in 1:5 and 1:6.69 K-β-Al$_2$O$_3$ compositions is higher as compared to 1:9 K-β-Al$_2$O$_3$ compositions. Therefore, 1:9 K-β-Al$_2$O$_3$ composition sintered at 1873 K for 30 min has been selected for the fabrication of composite electrolyte tubes used in the construction of antimony sensor tested in the present investigation. Figure 4, shows that conductivity in all the three compositions increases with increasing temperature and was found to be highest in 1:5 K-β-Al$_2$O$_3$ composition, this may also be due to the higher number of K$^+$ mobile ions present in the 1:5 K-β-Al$_2$O$_3$ composition as compared to 1:6.69 and 1:9 K-β-Al$_2$O$_3$ compositions.

Antimony sensor for molten lead has been designed and tested successfully. The antimony sensor was found to respond instantaneously with variation in antimony level in the Pb-Sb alloy, and attained stable EMF values in about 30 min of melting the alloy and EMF remains stable for about 15 min. The variation of EMF of the antimony sensor as a function of the natural logarithm of mole fraction of antimony in lead at 923 K was found to be Nerstian as shown in Figure 5 and tabulated in Table IV.

CONCLUSIONS

Antimony sensor for molten Pb-Sb alloy has been designed and tested successfully at 923 K. A novel antimony ion conducting composite solid electrolyte material has been synthesised (a heterogeneous mixture of K-β-Al$_2$O$_3$ and KSbO$_3$) and its tubes has been prepared and used for the fabrication of antimony sensor. The measured EMF of the antimony sensor as a function of varying antimony level in the Pb-Sb alloy indicates that sensor responded instantaneously. EMF of the antimony sensor was found to decrease as a function of increasing antimony level in the Pb-Sb alloy. A linear relationship between EMF and natural logarithm of mole fraction of antimony in lead was observed, indicating that sensor follows the Nerstian behaviour.

ACKNOWLEDGMENTS

We wish to acknowledge the Engineering and Physical Sciences Research Council (EPSRC) and Britannia Refined Metals Limited (BRML) for funding this research project. Technical discussions and assistance received during the course of this work from the BRML staff is highly appreciated.

REFERENCES

[1] G. M. Kale and R. Kurchania, "On-line Electrochemical Sensors in Molten Metal Processing Technology: A Review", Ceramic Transactions, The American Ceramic Society Inc., vol. 92, pp. 195-220, 1999.

[2] R. Kurchania and G. M. Kale, "Oxygen Potential in Molten Tin and Gibbs Energy of Formation of SnO_2 Employing an Oxygen Sensor", J. Mater. Res., vol. 15, pp. 1576-1582, 2000.

[3] R. Kurchania and G. M. Kale, "Measurement of Oxygen Potentials in Ag-Pb System Employing Oxygen Sensor", Metallurgical and Materials Transactions B, vol. 32B, pp. 417-421, 2001.

[4] G. M. Kale, A. J. Davidson and D. J. Fray, " Solid-state Sensor for Measuring Antimony in Non-ferrous Metals", Solid State Ionics, vol. 86-88, pp. 1101-1105, 1996.

[5] J. W. Fergus and S. Hui, "Solid Electrolyte Based Galvanic Cell for Measuring the Antimony Concentration in Molten Zinc", J. Electrochem. Soc., vol.143, pp. 2498-2502, 1996.

[6] Powder Diffraction File, Card No. 21-0618, Joint Committee on Powder Diffraction Standards, Swarthmore, PA (1979)

[7] Powder Diffraction File, Card No. 31-1034, Joint Committee on Powder Diffraction Standards, Swarthmore, PA (1979)

[8] Powder Diffraction File, Card No. 35-0664, Joint Committee on Powder Diffraction Standards, Swarthmore, PA (1979)

PREPARATION AND CHARACTERIZATION OF IRON OXIDE-ZIRCONIA NANO POWDER FOR ITS USE AS AN ETHANOL SENSOR MATERIAL

C V Gopal Reddy* and S A Akbar
CISM, Dept. of Materials Science & Eng.
The Ohio State University
Columbus, OH 43210
USA.
*e-mail: chada@cism.ohio-state.edu

W Cao, O K Tan and, W Zhu
Sensors & Actuators Lab
School of EEE
Nanyang Technological University
Singapore-639798

ABSTRACT

Powders of composition $xZrO_2$-$(1-x)Fe_2O_3$ were prepared by several methods such as high-energy ball milling, co-precipitation and hydrazine methods. This paper presents the effect of the preparation methods and annealing temperatures on the ethanol gas sensitivity. Noble metals such as Pt and Pd were added in order to examine the effects on ethanol gas sensitivity. Sensors based on 1wt.% Pt + $xZrO_2$- $(1-x)Fe_2O_3$ demonstrated an excellent sensing performance at 230 °C for 1000 ppm of ethanol. The gas-sensing behavior of these materials to various reducing gases like CO, CH_4 and H_2 was also studied. This sensor showed good selectivity toward ethanol and thus could effectively be used as a breath sensor.

INTRODUCTION

Nano materials is a broad and interdisciplinary area of research that has seen tremendous growth in the past several years. Recent interest in nano-structured materials has been stimulated by the work of Gleiter[1] on materials produced by the gas condensation method. The method of high-energy ball milling has also received significant attention in recent years for nano particle formation. The nano-crystalline materials are usually considered to be of meta-stable nature due to high interfacial and grain boundary energies. Hence their structure and properties depend on the preparation method, as well as the time-temperature history. They show distinct properties due to different atomic structures in the interfacial region.[2] When the size of the crystals becomes smaller than a critical value, the property changes can be dramatic leading to the possibility of creating materials with unusual functionality engineered through size control.[3] Nano-crystals have been prepared for many years by chemical means such as sol-gel,

---

To the extent authorized under the laws of the United States of America, all copyright interests in this publication are the property of The American Ceramic Society. Any duplication, reproduction, or republication of this publication or any part thereof, without the express written consent of The American Ceramic Society or fee paid to the Copyright Clearance Center, is prohibited.

hydrothermal, and precipitation techniques.[4,5] The size of the grains in a nano-structured material has pronounced effects on many of its properties, the best known being the increase in strength and hardness. This dependence of properties on grain size is of utmost importance in the control of material synthesis and processing.

Fine particles of ferrites have been of interest due to their applications in the preparation of high-density ferrite cores, as suspension materials in ferromagnetic liquids, and as catalysts. Among these, iron oxides in particular are technologically useful as pigments and semiconductors, and also for their magnetic properties.[6] The stability and semiconducting properties of α-Fe_2O_3 allows it to be used as a photo catalyst.[7,8] Semiconducting thick films of Fe_2O_3 have been studied earlier as a sensor for CH_4, H_2 and NH_3.[9] Cantalini et al.[10,11] have also reported α-Fe_2O_3 based gas sensors. Dong et al.[12] have prepared nano-sized particles of pure γ-Fe_2O_3 by H_2-Ar arc plasma. Thick films of these sensors were shown to be sensitive to NH_3. Zeng et al. have studied high stability γ-Fe_2O_3 as a possible gas sensor for detecting several reducing gases.[13] SnO_2-Fe_2O_3 thin films have also been reported as effective humidity sensor.[14] Some researchers also reported tin doped α-Fe_2O_3 systems for sensing CH_4, CO gases.[15,16] This paper reports sensing performance of $xZrO_2$-$(1-x)Fe_2O_3$ systems.

EXPERIMENTAL

Zirconia (0.1 M) doped iron oxide (0.9 M) was prepared by high-energy ball milling from a mixture of hematite (α-Fe_2O_3) (Alfa, 99.9 %) and zirconia (m-ZrO_2) (Alfa, 99.9 %) as explained in our earlier paper.[17] In the co-precipitation method, 0.9M iron nitrate, $(Fe(NO_3)_3 \cdot 9H_2O$ (Merck, 99 %) and 0.1M zirconium tetra chloride, $ZrCl_4$ (Aldrich, 99 %) were taken and dissolved in deionized water (18 MΩ). Ammonia (25 %) solution was added slowly drop by drop to the above mixture solution with continuous stirring. The pH value was adjusted to around 7.5 so as to obtain a complete precipitation. The stirring was then stopped and the precipitate was allowed to settle. The products were removed from the filtration and washed with deionized water until no nitrate and chloride ions were detected by a brown ring test and silver nitrate solution. The products were dried in an oven at 120 °C for overnight.

In the hydrazine method, 0.9M ferric nitrate nano hydrate $(Fe(NO_3)_3 \cdot 9H_2O)$, 0.1M zirconium tetra chloride ($ZrCl_4$) and hydrazine mono hydrate $((NH_2)_2H_2O)$ (TCI, 98%) were used as the starting materials. Ferric nitrate and zirconium tetra chloride solution were prepared by dissolving above amounts in 0.1M of deionized water (18 MΩ). This aqueous solution (pH=1.86) was poured in a beaker and was stirred on a magnetic stirrer at 60 °C for 30 minutes. Hydrazine monohydrate 0.8M was added slowly to the nitrate solution through a pipette by maintaining a constant stirring until the resulting precipitate reached to pH= 8.[18]

The precipitate was separated by filtering, washed more than 15 times in deionized water to remove the adsorbed hydrazine, chloride and nitrate ions, and then dried in air. The details of the preparation procedure for these three routes are shown in Figure 1.

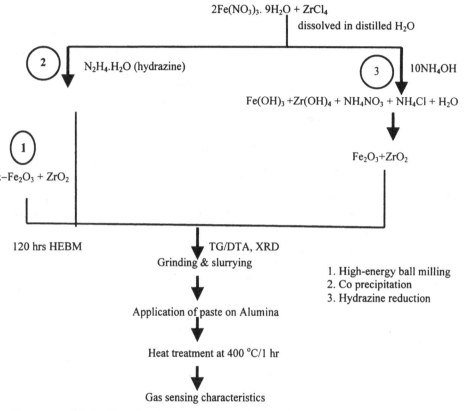

Figure 1. Flow-chart for the preparation of the sensor device: (1) high energy ball milling; (2) co-precipitation and (3) hydrazine reduction.

Thoroughly washed and dried material was then subjected to thermal gravimetric-differential thermal analysis (TGA/DTA) and X-ray diffraction (XRD). Perkin-Elemer TGA-7 and DTA-7 were used to measure the thermal properties of the materials. TG/DTA was conducted in air at a heating rate of 10 °C/min in the temperature range from the room temperature (RT) up to 1000 °C; α-Al_2O_3 was used as the reference. These materials were studied by using a Rigaku X-ray diffractometer (CuK$_\alpha$ radiation, λ=1.5406 Å). Pastes of these materials were prepared using a commercial organic vehicle 400 (from ESL) and screen-printed onto Al_2O_3 substrates with inter-digital Au electrodes. The samples were then annealed at temperatures from 400 °C to 600 °C for an hour in air. The gas sensing properties were measured by using a computer-controlled gas sensing characterization system.

RESULTS AND DISCUSSION

Characterization of Material

Figure 2(a) shows the TGA-DTA curves of as-prepared $xZrO_2\text{-}(1\text{-}x)Fe_2O_3$ material by high energy ball-milling method. There is a small weight loss (loss of surface water) observed from RT up to about 850 °C. At 850 °C, a significant weight loss was observed with a total loss of about 8 %. The DTA data shows a small endothermic peak around 850 °C consistent with TGA data.

The TGA data of as prepared powder by the co-precipitation method shows a total loss of 9 % on heating (Figure 2b). The weight loss amounting to 3.5% from RT up to 100 °C is due to the loss of surface water. The rest of the weight loss up to 300 °C is due to the loss of hydroxyl groups from the powder. The DTA data shows the associated endothermic peaks at 100 °C and 255 °C, respectively. It also shows a sharp exothermic peak at 340 °C, which is attributed to a change from the amorphous to the crystalline phase.

The TG-DTA data of $xZrO_2\text{-}(1\text{-}x)Fe_2O_3$ prepared from hydrazine is shown in Figure 2(c). An endothermic peak appears at 103 °C representing the loss of surface water. An exothermic peak appears at 273 °C that corresponds to a phase change from Fe_2O_3 (Maghemite-Q) to $\alpha\text{-}Fe_2O_3$. Note that the endothermic peak associated with the loss of hydroxyl group is missing because this method directly produces oxides. Another small exothermic peak observed at 725 °C without any associated change in the weight loss may be due to phase transformation in ZrO_2.

The XRD patterns for the mechanically alloyed $xZrO_2\text{-}(1\text{-}x)Fe_2O_3$ samples are shown in Figure 3a, for x = 0.1. All the peaks match with the hematite ($\alpha\text{-}Fe_2O_3$) phase. With the increase of milling time, the peaks broaden due to the decrease in the grain size. For particles milled for 2 hrs, the particle size reduced down to about 25 nm. From the XRD peaks of 120 hrs milled powders, the average crystallite size calculated using Scherrer formula[19] is about 8 nm.

The XRD patterns of $Fe_2O_3\text{-}ZrO_2$ prepared by the co-precipitation method and calcined at 120 °C for 12 hrs, 350 °C 2 hrs. and 500 °C for 2 hrs are shown in Figure 3b. While the sample calcined at 120 °C shows an amorphous behavior, samples calcined above 350 °C exhibit a crystalline phase with peaks matching with the reported phase of hematite (JCPDS card No. 33-664). From the XRD peaks, the estimated average crystallite size is ~ 177.3Å. Figure 3 (c) shows XRD patterns of $xZrO_2\text{-}(1\text{-}x)Fe_2O_3$ that are treated at different temperatures. From RT up to 300 °C, it shows only the Maghemite-Q phase with tetragonal structure (Card No. 25-1402). Increase in the calcination temperature above 300 °C leads to the appearance of the α-Fe$_2$O$_3$ peaks. The average crystallite size calculated using Scherrer formula is about 82 nm.

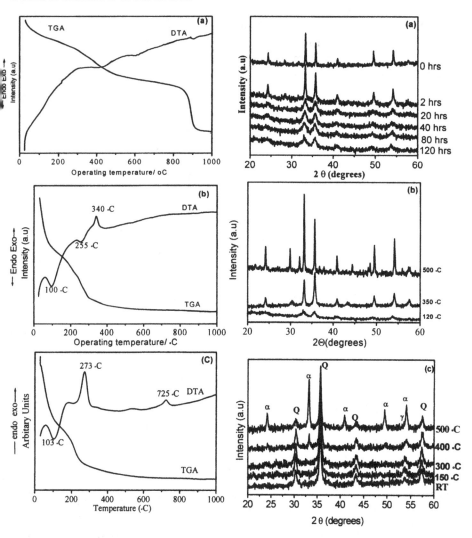

Figure 2 TGA-DTA data of xZrO$_2$-(1-x)Fe$_2$O$_3$ prepared by: (a) high-energy ball milling; (b) co-precipitation; and hydrazine reduction method.

Figure 3 XRD patterns of xZrO$_2$-(1-x)Fe$_2$O$_3$ prepared by: (a) high-energy ball milling; (b) co-precipitation; and (c) hydrazine method. (α: α-Fe$_2$O$_3$, γ: γ-Fe$_2$O$_3$ and Q: Maghemite).

Gas Sensing Characteristics of High-Energy Ball Milling Material

The gas sensitivity (S) is defined as the ratio of the resistance of the sensor in air (R_a) to that in the test gas (R_g). We first examined the effect of different weight percentages of ZrO_2 on gas sensing properties. It was found that 10 wt.% ZrO_2 doped Fe_2O_3 sensor had a much better sensitivity to ethanol gas than a single phase Fe_2O_3 based sensor. Hence, all the sensing measurements were conducted on samples with x = 0.1.

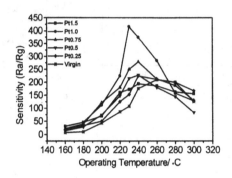

Figure 4 Sensitivity vs. the operating temperature of x ZrO_2-(1-x) Fe_2O_3 for vargin and Pt impregnated samples by high-energy ball milling.

Figure 4 shows the sensitivity of the virgin and different weight percentages of Pt impregnated material as a function of the operating temperature. Corresponding amounts of Pt chlorides were impregnated in x ZrO_2-(1-x)Fe_2O_3 followed by high-energy ball milling as described above. This device was annealed at 400 °C for 1 hr. before sensing measurements. The sample with 1 wt.% Pt showed the maximum sensitivity (~ 416) for 1000 ppm ethanol vapor at 230 °C. We also tried Pd impregnated samples, but Pt gave better response. Hence, for further studies, we selected the 1wt. % Pt sample.

Figure 5 shows the sensitivity versus the operating temperature for different milling hours from 40 to 120. The sensitivity increases with increase in milling time since longer milling produces finer particles providing higher surface area for gas sensing. For the same reason, samples calcined at lower temperatures showed better sensitivity. For these samples, the optimum operating temperature was found to be 230 °C.

Figure 6 shows the sensitivity of xZrO_2–(1-x)Fe_2O_3 to various test gases as a function of the operating temperature. It is seen that the sensor exhibits good selectivity by sensing only ethanol vapor at an operating temperature of 230 °C in the presence of other interfering gases like CH_4, CO and H_2.

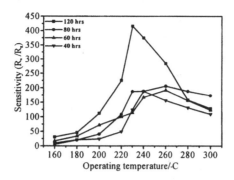

Figure 5 Sensitivity vs. operating temperature of x ZrO_2-(1-x) Fe_2O_3 + 1 wt% Pt for different milling hours.

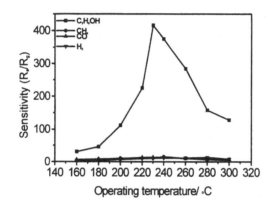

Figure 6 Sensitivity vs. operating temperature for x ZrO_2-(1-x) Fe_2O_3 + 1 wt% Pt with different reducing gases.

Gas Sensing Characteristics of $xZrO_2$–$(1-x)\alpha$-Fe_2O_3 Prepared by the Co-Precipitation Method

Figure 7 shows sensitivity versus the operating temperature of virgin and 1 wt.% Pt impregnated ZrO_2–(1-x)Fe_2O_3 material prepared by the co-precipitation method. For the virgin sample the maximum sensitivity is around 170, but the Pt sample shows a maximum value of 486 at 230 °C. So we optimized the composition at 1 wt% Pt and varied the annealing temperature. Figure 8 shows the sensitivity versus the operating temperature for samples annealed at different temperatures from 400 °C to 600 °C. It is clearly seen that there is not much difference in sensitivity with the annealing temperature, presumably because the crystallite size doesn't change.

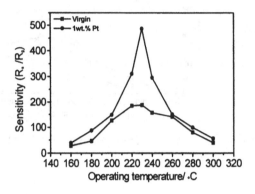

Figure 7 Sensitivity vs. operating temperature of x ZrO_2-(1-x) Fe_2O_3 + 1 wt% Pt prepared by co-precipitation for 1000 ppm ethanol.

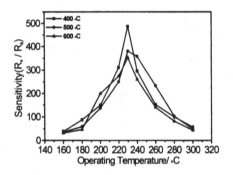

Figure 8 Sensitivity to 1000 ppm ethanol as a function of the operating temperature for xZrO_2-(1-x)Fe_2O_3 + 1wt.% Pt (co-precipitation), annealed at 400, 500 and 600 °C for 1 hour.

Gas Sensing Characteristics of xZrO$_2$–(1-x)Fe$_2$O$_3$ Prepared by Hydrazine Method

Figure 9 shows the sensitivity versus the operating temperature of virgin and 1 wt.% Pt impregnated ZrO_2–(1-x) Fe_2O_3 material prepared from the hydrazine method. It is clearly observed that there is not much difference in sensitivity between virgin and Pt incorporated sample. Here we observed a maximum sensitivity of 600 at 230 °C for virgin and 800 for the Pt impregnated sample.

These values are much higher compared to those of the ball milling and co-precipitation methods.

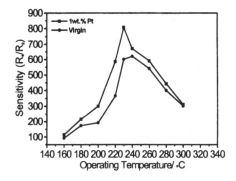

Figure 9 Sensitivity vs. operating temperature of x ZrO_2-(1-x) Fe_2O_3 + 1 wt% Pt prepared by hydrazine reduction for 1000 ppm ethanol.

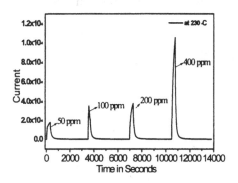

Figure 10 Current vs. time response of sensors prepared hydrazine reduction for different concentrations of ethanol at 230 °C.

Figure 10 shows the relationship between the sensitivity and the ethanol vapor concentration for the sensor operating at 230 °C. The sensitivity for 50 ppm ethanol is as high as 160. It was also observed that the sensor senses even less than 10 ppm with a substantial sensitivity. Figure 11 shows the sensitivity versus time for 1000 ppm ethanol showing a very fast response time and a sluggish recovery. Finally, all the three methods for ethanol sensing are compared in a single graph as shown in Figure 12. From this graph we can conclude that $xZrO_2$-$(1-x)Fe_2O_3$ prepared by the hydrazine method gives the best results as an ethanol sensor compared to the other methods.

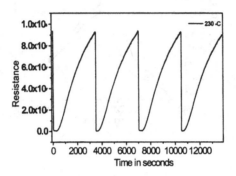

Figure 11 The response and recovery behavior of the $xZrO_2$-$(1-x)Fe_2O_3$+ 1 wt% Pt (hydrazine reduction) sensor to 1000 ppm ethanol at 230 °C.

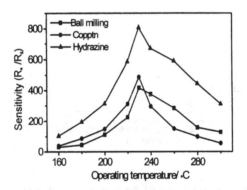

Figure 12 Comparison of the sensing behavior of x ZrO_2-$(1-x)Fe_2O_3$ + 1 wt. % Pt in 1000 ppm ethanol.

CONCLUSIONS

Nano powders of $xZrO_2$-$(1-x)Fe_2O_3$ prepared from hydrazine reduction method produces maghemite-Q (Fe_2O_3) which shows better sensitivity towards ethanol. The doping of Pt (1.0 wt%) produces a remarkable improvement in sensitivity. It shows good sensitivity even at < 50 ppm with sufficient sensitivity. It is also highly selective compared to other interfering gases like CH_4, CO and H_2. The maximum sensitivity of $R_a/R_g \sim 800$ is observed for sensor annealed at 400 °C for 1 hr and an operating temperature of 230 °C.

REFERENCES

[1] H. Gleiter, "Nanocrystalline Materials," *Progress in Materials Science,* **33** (4) 223-315 (1989).

[2] H. Gleiter, "Nanostructured materials: State of the art and perspectives," *Nanostructured Materials,* **6** 3-14(1995).

[3] R.W. Siegel, "Creating Nanophase Materials," *Scientific American,* **275** [6] 74-79 (1996).

[4] Y. Murase and E. Kato, "Thermal Changes in Texture of Aggregates of Ultra-Fine Crystallites in Hydrolysed Zirconia Particles," *Journal of Crystal Growth* **50** (1980) 509.

[5] S. Komarneni, E. Fregeau, E. Breval and R. Roy, "Hydrothermal Preparation Of Ultrafine Ferrites and their Sintering," *Journal of The American Ceramic Society,* **71** [1] C26-C28 (1988).

[6] R.M. Cornell and U. Schwertmann, "The Iron Oxides," pp. 115, VCH, New York, 1996.

[7] B.L.Yang and H.H. Kung, Oxygen on Iron-Oxide-Effect on the Selective Oxidation of Butane," Journal of Catalysis, **77** (2) 410-420 (1982).

[8] J.K. Leland and A.J. Bard, "Photochemistry of Colloidal Semiconducting Iron-Oxide Polymorphs," *Journal of Physical Chemistry,* **91** [19] 5076-5083(1987).

[9] V.V. Malyshev, A.V. Eryshkin, E.A. Koltypin, A.E. Varfolomeev and A.A. Vasiliev, "Gas sensitivity of semiconductor Fe_2O_3 based thick-film sensors to CH_4, H_2 and NH_3," *Sensors and Actuators B,* **19** 434-436 (1994).

[10] C. Cantalini, M. Faccio, G. Ferri and M. Pelino, "Microstructural and electrical properties of Si-doped- α-Fe_2O_3 sensor," *Sensors and Actuators B,* **16** 293-298 (1993).

[11] C. Cantalini, M. Faccio, G. Ferri and M. Pelino, "The influence of water vapour on carbon monoxide sensitivity of α-Fe_2O_3 microporous ceramic sensors," *Sensors and Actuators B,* **18-19** 437-442 (1994).

[12] L.F. Dong, Z. Lui and Z. Zhang, "Gas sensitivity of rare earth doped α-Fe_2O_3," *Gongneng Cailia,* **26** 416-417 (1995).

[13]H. Zeng, H. Luo, T. Zhang and R. Zhang, "Synthesis of high stability γ-Fe_2O_3 and its gas sensing characteristics," Gongneng Cailia, **26** [4] 305-308(1995).

[14]T.M. Racheva, O. Stamb, I.T. Lova and T. Donchev, "Humidity Sensitive Characteristics of SnO_2-Fe_2O_3 Thin Films Prepared by Spray Pyrolysis," Journal of Materials Science, **29** [1] 281-284 (1994).

[15]M. Takano, Y. Bando N. Nakanishi, M. Sakai and H. Okinaka, "Characterization of Fine Particles of the α-Fe_2O_3-SnO_2 System with Residual So_4^{2-} Ions on the Surface," Journal of Solid State Chemistry, **68** [1] 153-162 (1987).

[16]P. Bonzi, L.E. Depero, F. Parmigiani, C. Perego, G. Sberveglieri and G. Quattroni, "Formation and Structure of Tin-Iron Oxide Thin-Film CO Sensors," Journal of Material Research, **9** [5] 1250-1256 (1994).

[17]O.K. Tan, W. Zhu and J.Z. Jiang, "Nano structure and $xSnO_2$-$(1-x)Fe_2O_3$ material preparation for good gas sensing properties," Proc. of 4th natl Symp. on progress of mat. Res., singapore, pp 542- 545 (1998).

[18]C.V. Gopal Reddy, K. Kalyana Seela and S.V. Manorama, "Preparation of γ-Fe_2O_3 by the Hydrazine method: Application as an alcohol Sensor," International Journal of Inorganic materials, **2/4** 301-307(2000).

[19]H.P. Klug and L.E. Alexander, "X-ray diffraction procedure for polycrystalline materials," pp 125, John Wiley & Sons, New York, 1974.

SYNTHESIS AND CHARACTERIZATION OF 2-3 SPINELS AS MATERIAL FOR METHANE SENSOR

S. Poomiapiradee and R.M.D. Brydson
Department of Materials
School of Process, Environmental and Materials Engineering
University of Leeds
Leeds LS2 9JT
United Kingdom

G.M. Kale
Department of Mining and Mineral Engineering,
School of Process, Environmental and Materials Engineering
University of Leeds
Leeds LS2 9JT
United Kingdom

ABSTRACT

$NiFe_{2-x}Al_xO_4$ and $NiFe_{2-x}Cr_xO_4$, where $0 \leq x \leq 2$, were synthesized by intimately mixing the fine powders of NiO, Fe_2O_3 and γ-Al_2O_3; and NiO, Fe_2O_3 and Cr_2O_3, respectively, in accordance with the formula of the composition. Then those powders were calcined in closed alumina crucibles at various temperatures ranging from 800 to 1500 °C for 3 hours. The calcined powders were characterized by an X-ray diffractometer. The values of "a" (lattice parameter) were calculated from the XRD pattern. Density and electrical conductivity of samples was also measured. The response of the sensing materials in gaseous atmosphere (2.5 % CH_4 in air + air) over a range of methane concentration has been measured at 180 °C. The results of the experimental investigation are presented in the following section.

INTRODUCTION

Although methane (CH_4) emission from various sources such as landfill sites, coal mines, gas leakage, oil and gas production, offshore wells, and domestic fuel, has decreased by 16 %[1] since 1970, it is still about 4.5 million tons per annum in UK alone. Therefore, it is important to design a sensor for continuous monitoring of CH_4 at various sources.

Gas present in the environment may be harmless, toxic or flammable.

To the extent authorized under the laws of the United States of America, all copyright interests in this publication are the property of The American Ceramic Society. Any duplication, reproduction, or republication of this publication or any part thereof, without the express written consent of The American Ceramic Society or fee paid to the Copyright Clearance Center, is prohibited.

Leakage of toxic and flammable gases, therefore, may cause serious problems for health and safety of human life. Some inflammable gases have very low explosive limit; for example CH_4 has lower explosive limit (LEL) 5.0 vol. %[2], which can easily cause fire if it leaks. For this reason, on-line gas measuring sensor system is necessary for inspection and monitoring tasks in the modern oil, gas and chemical-based industries.

Over the past few decades alternative sensors for methane have been developed using different materials and fabrication processes. The gas-sensing properties of semiconducting oxide films, namely ZnO and SnO_2 have been studied for a long time and SnO_2 gas sensors have been available for more than 20 years.[3] Other base materials used for methane detection are Ga_2O_3 thin films and $CaZrO_3$.[4,5,6]

In the ceramic industry, a variety of sensor elements have been developed. Many solid state gas sensors used to measure various gases (e.g. O_2, CO_2, H_2, CH_4) are based on the interaction between an electrical system (the sensing element) and a chemical system (the measurement gas).[7] For many years, gas sensors based on resistive changes of selected semiconductor materials have been successfully used. They are used as domestic gas detectors to trigger an alarm at a specific gas concentration as prevention measure against gas explosion hazards in households and in industry. Metallic oxides, such as SnO_2 and ZnO, are generally used as base materials, due to their surface properties. They are very suitable for the detection of toxic gases (e.g. CO, H_2S, NO_2) and inflammable gases (e.g. CH_4, H_2). However, these simple binary oxides are inherently non-selective and are strongly affected by atmospheric contaminants such as moisture.[8] Therefore, an empirical development of alternative sensor materials for methane has been studied in order to develop a methane sensor material, which would enhance selectivity or sensitivity and overcome problems of moisture interference. Moreover, it may indirectly reduce gas explosion hazards caused by methane, especially around coal mines and landfill sites due to subsurface methane migration.

EXPERIMENTAL PROCEDURES

Powder Preparation

Nickel ferrite ($NiFe_2O_4$) was synthesized by intimately mixing the fine powders of NiO and Fe_2O_3 in equimolar ratio under isopropanol in a polypropylene ball mill with alumina balls for 6 hours drying the slurry under an IR lamp using magnetic stirrer. The dried, milled powders were sieved through a 300 μm mesh size polyester sieve and finally calcining the mixed powders in a closed alumina crucible at various temperatures ranging from 800 to 1400 °C for 3 hours. Nickel aluminate ($NiAl_2O_4$) was prepared from an intimate mixture of NiO and γ-Al_2O_3 in a molar ratio of 1:1 following the same procedure as $NiFe_2O_4$ and eventually calcining the mixed oxide powders in a closed alumina crucible at different temperatures ranging from 800 to 1500 °C for 3 hours. Nickel chromite ($NiCr_2O_4$) was synthesized by calcining a mixture of NiO and Cr_2O_3 in the desired proportions at various temperatures ranging from 800 to 1500 °C for 3 hours. Solid solutions of $NiFe_{2-x}Al_xO_4$ and $NiFe_{2-x}Cr_xO_4$ were prepared by the solid-state reaction of NiO, Fe_2O_3 and γ-Al_2O_3; and NiO, Fe_2O_3 and Cr_2O_3, respectively. The composition of $NiFe_{2-x}Al_xO_4$ samples studied were $x = 0.1, 0.2, 0.3, 0.4, 0.5, 1$ and 1.5, whereas the composition of $NiFe_{2-x}Cr_xO_4$ samples were $x = 0.25, 0.75, 1.25$ and 1.50. The oxides of appropriate proportions were mixed uniformly following the same procedure as $NiFe_2O_4$ and ultimately calcined the mixed powders in a closed alumina crucible at various temperatures ranging from 1200 to 1500 °C for 3 hours.

Sensing Materials Preparation

The powder of each spinel-type oxide and their solid solutions was die pressed uniaxially into a pellet of approximately 1.2 cm diameter at a pressure of 150 MPa, and then sintered at various temperatures ranging from 1200 to 1600 °C for 2 hours. The sintered pellet, which was coated with platinum paint to make blocking electrodes and platinized at 900 °C for 30 minutes in order to get rid of the binders in the platinum paint, which may affect the results of electrical conductivity. The pellet was mounted in a spring-loaded quartz rig with electrical contacts being made by means of Pt gauze (Aldrich, 99.9 %, 52 mesh) spot welded to Pt wires (Aldrich, 99.99 %, 0.127 mm dia.).

Characterization

Phase characterization: Powders of $NiFe_2O_4$, $NiAl_2O_4$, $NiCr_2O_4$, $NiFe_{2-x}Al_xO_4$ ($x = 0.1, 0.2, 0.3, 0.4, 0.5, 1, 1.5$) and $NiFe_{2-x}Cr_xO_4$ ($x = 0.25, 0.75, 1.25, 1.50$) phases were characterized by an X-ray diffractometer (Philips APD 1700) at room temperature. $NiFe_2O_4$ powders were scanned for a 2θ-range from 25-65°, whereas $NiAl_2O_4$ was scanned from 15-70°. $NiCr_2O_4$ powders were scanned for a 2θ-range from 10-73°. The solid solutions of $NiFe_{2-x}Al_xO$ were scanned from 15-71°, whereas $NiFe_{2-x}Cr_xO_4$ was scanned from 10-73°. The type of scan used was continuous scan, with a step size of 0.10 degrees two theta (°2θ) and scan speed of 0.05 °2θ s^{-1}, using Cu Kα radiation.

Determination of lattice parameters of solid solutions by using Nelson and Riley's Extrapolation against ½ ($cos^2 \theta/sin\theta + cos^2\theta/\theta$)[9,10]: From XRD pattern at 1400 °C of each solid solution, the values of "a" for each peak can be calculated by the following equation:

$$\frac{1}{d^2} = \frac{h^2 + k^2 + l^2}{a^2} \quad \text{(for cubic structure)}$$

where $d_{(hkl)}$ is the interplanar spacings of the planes (hkl) and

"a" is the lattice parameter of the crystalline unit cell

Then a graph between the values of "a" and $1/2\ (\cos^2\theta/\sin\theta + \cos^2\theta/\theta)$ was plotted for each solid solution. The lattice constants of each solid solution were determined by extrapolation technique.

Density measurement of pellets: Density of sintered pellets was measured by gas displacement pycnometer (AccuPyc 1330 Pycnometer) using He gas. Samples were dried very thoroughly in an oven set at approximately 110 °C for 48 hours and then cooled down to ambient temperature in a desiccator prior to density measurement.

RESULTS AND DISCUSSION

XRD Analysis

$NiFe_2O_4$, $NiAl_2O_4$ and $NiCr_2O_4$ crystallize in a cubic spinel structure in which Ni^{2+}, Fe^{3+}, Al^{3+} and Cr^{3+} cations are distributed in tetrahedral and octahedral sites of a face-centered cubic lattice of oxygen ion having an *Fd3m* symmetry. There are 64 tetrahedral sites and 32 octahedral sites. In the case of normal 2-3 spinel, all the divalent cations occupy 1/8 of tetrahedral sites and all the trivalent cations occupy half of octahedral sites. In the case of inverse spinel, half of the trivalent cations occupy 1/8 of the tetrahedral sites, whereas all the divalent cations and the remaining half of the trivalent cations occupy half of the octahedral sites. The intermediate cation distribution between the normal and the inverse spinel depends on size effect, Madelung constant, crystal field effect, Jahn-Teller distortion, covalency effect and the thermodynamics of cation mixing on crystallographically non-equivalent sites of spinel lattice. $NiFe_2O_4$ crystallizes in an inverse spinel structure, whereas $NiAl_2O_4$ crystallizes in a partially inverse spinel structure. The octahedral site preference energy of Cr^{3+} cation[11] is significantly higher than Ni^{2+}, Fe^{3+} and Al^{3+} and hence $NiCr_2O_4$ crystallizes in a normal spinel structure. The cation distribution in $NiFe_2O_4$ can be written as $(Fe^{3+})_{tet}[Ni^{2+}Fe^{3+}]_{oct}O_4$ and that of $NiAl_2O_4$ can be written as $(Ni_{0.2}^{2+}Al_{0.8}^{3+})_{tet}[Ni_{0.8}^{2+}Al_{1.2}^{3+}]_{oct}O_4$. Cation distribution of $NiCr_2O_4$ can be written as $(Ni^{2+})_{tet}[Cr_2^{3+}]_{oct}O_4$.

$NiFe_2O_4$ pattern after calcining the mixtures of NiO and Fe_2O_3, figure 1 showed that $NiFe_2O_4$ was formed at 800 °C but traces of NiO and Fe_2O_3 were still detected by XRD. At 1000 °C $NiFe_2O_4$ gradually developed its formation and after calcining at 1200 °C for 3 hours, a completely developed $NiFe_2O_4$ phase was presented, and no traces of unreacted NiO or Fe_2O_3 were detected from the X-ray diffraction pattern. NiO reacted with γ-Al_2O_3 to form $NiAl_2O_4$ completely at 1400 °C for 3 hours, as shown in figure 2. At 1200 °C, traces of NiO and α-Al_2O_3 were detected by XRD and $NiAl_2O_4$ formation started at this temperature. Figure 3 indicated that the complete formation of single-phase $NiCr_2O_4$ occurred at 1400 °C. No traces of unreacted NiO or Cr_2O_3 were detected from the XRD pattern.

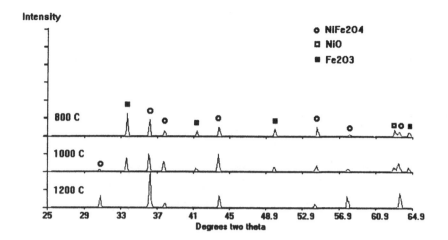

Figure 1. XRD patterns of NiFe$_2$O$_4$ powder calcined at different temperatures.

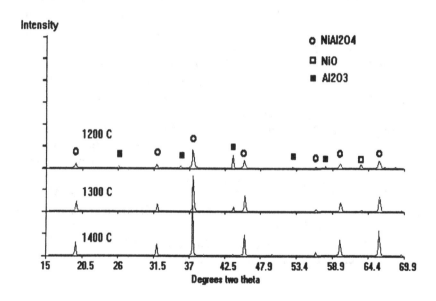

Figure 2. XRD patterns of NiAl$_2$O$_4$ powder calcined at different temperatures.

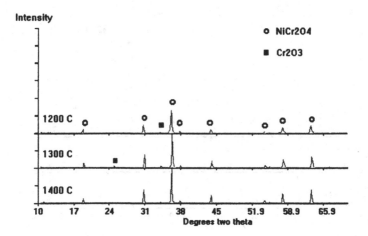

Figure 3. XRD patterns of NiCr$_2$O$_4$ powder calcined at different temperatures.

Solid solutions (NiFe$_{2-x}$Al$_x$O$_4$, where x = 0.1, 0.2, 0.3, 0.4, 0.5, 1.0, and 1.5) patterns after calcining the mixtures of NiO, Fe$_2$O$_3$ and γ-Al$_2$O$_3$ confirmed that NiFe$_{2-x}$Al$_x$O$_4$ was formed at 1200 °C, but traces of NiO and α-Al$_2$O$_3$ were still detected by XRD. At 1300 °C NiFe$_{2-x}$Al$_x$O$_4$ peaks became more intense whereas the intensity of NiO and α-Al$_2$O$_3$ peaks decreased. After calcining at 1400 °C for 3 hours, NiFe$_{2-x}$Al$_x$O$_4$ solid solution was completely formed, and NiO and α-Al$_2$O$_3$ peaks were not detected. Solid solutions (NiFe$_{2-x}$Cr$_x$O$_4$, where x = 0.25, 0.75, 1.25, and 1.50) were completely formed at 1400 °C. No traces of NiO, Fe$_2$O$_3$ or Cr$_2$O$_3$ were detected.

In comparison to the reaction of NiO and Fe$_2$O$_3$ to form NiFe$_2$O$_4$, the formations of NiAl$_2$O$_4$ from NiO and Al$_2$O$_3$; and NiCr$_2$O$_4$ from NiO and Cr$_2$O$_3$ were relative slow. This is mainly due to the excess thermodynamic stability of one of the reactants i.e. Al$_2$O$_3$ and Cr$_2$O$_3$ relative to Fe$_2$O$_3$. For instance, the free energy of formation of Fe$_2$O$_3$ from its element is -437,584 Jmol^{-1} at 1500 K[12], whereas that of α-Al$_2$O$_3$ is -1,196,519 Jmol^{-1} and that of Cr$_2$O$_3$ is -749,380 Jmol^{-1} at the same temperature. For the same reason, the kinetic of formation of NiFe$_{2-x}$Al$_x$O$_4$ and NiFe$_{2-x}$Cr$_x$O$_4$ spinel solid solutions progressively decreased as x increased from zero to two.

Solid Solution Lattice Parameters

From Table I and II, it can be concluded that the lattice constant (a) decreases with increasing the aluminium and chromium content, respectively. From Bragg's Law and the d-spacing formulae, it can be deduced that a decrease in the unit cell parameters causes a decrease in the d-spacings of the powder lines. Therefore, the entire XRD pattern of the solid solution is shifted to higher values of 2θ, when aluminium or chromium contents increased.

To the best of our knowledge, there is no information in the literature on the variation of lattice parameters as a function of composition of solid solutions for the system $NiFe_{2-x}Al_xO_4$, $0 \leq x \leq 2$ and $NiFe_{2-x}Cr_xO_4$, $0 \leq x \leq 2$. For $NiFe_{2-x}Al_xO_4$ system, nine different compositions were prepared ranging from $x = 0$ to $x = 2$, whereas 6 of those were prepared at the same range of the spinel solid solution system $NiFe_{2-x}Cr_xO_4$. The lattice parameters (a) of each composition of the cubic 2-3 spinel solid solution were calculated from the XRD patterns using Nelson and Riley's Extrapolation method. Figure 4(a) and 4(b) showed the variation of lattice parameters as a function of composition of the spinel solid solutions $NiFe_{2-x}Al_xO_4$ and $NiFe_{2-x}Cr_xO_4$ systems, respectively. The values of "a" calculated for each composition are shown in Table I and II. These are also compared with the theoretical lattice constant obtained using Vegard's Law approximation. The lattice constant for $NiFe_2O_4$, $NiAl_2O_4$ and $NiCr_2O_4$ were taken from JCPDS-ICDD.[13,14] It can be seen from figure 4 that the lattice parameters of solid solutions obtained from the present study show the negative departures from Vegard's Law and the lattice parameters decreased with increasing Al and Cr content of spinel solid solutions, respectively. This is due to the substitution of Fe^{3+} by Al^{3+} and Cr^{3+} respectively; both of which have ionic radius smaller than Fe^{3+}.[15]

Table I. Values of "a" for $NiFe_{2-x}Al_xO_4$ system (x = 0, 0.1, 0.2, 0.3, 0.4, 0.5, 1.0, 1.5 and 2.0) obtained from Extrapolations against 1/2 ($cos^2\theta/sin\theta$ + $cos^2\theta/\theta$).

Composition	Theoretical lattice constant, a_{th} (Å)	Experimental lattice constant, a_{exp} (Å)
$NiFe_2O_4$	8.339	8.343
$NiFe_{1.9}Al_{0.1}O_4$	8.325*	8.326
$NiFe_{1.8}Al_{0.2}O_4$	8.311*	8.306
$NiFe_{1.7}Al_{0.3}O_4$	8.296*	8.28
$NiFe_{1.6}Al_{0.4}O_4$	8.282*	8.267
$NiFe_{1.5}Al_{0.5}O_4$	8.267*	8.26
$NiFeAlO_4$	8.194*	8.16
$NiFe_{0.5}Al_{1.5}O_4$	8.122*	8.106
$NiAl_2O_4$	8.048	8.033

Note: * From Vegard's Law approximation

Table II. Value of "a" for NiFe$_{2-x}$Cr$_x$O$_4$ system (x = 0, 0.25, 0.75, 1.25, 1.50, and 2.00) obtained from Extrapolations against 1/2 ($\cos^2\theta/\sin\theta + \cos^2\theta/\theta$).

Composition	Theoretical lattice constant, a_{th} (Å)	Experimental lattice constant, a_{exp} (Å)
NiFe$_2$O$_4$	8.339	8.343
NiFe$_{1.75}$Cr$_{0.25}$O$_4$	8.336*	8.338
NiFe$_{1.25}$Cr$_{0.75}$O$_4$	8.330*	8.324
NiFe$_{0.75}$Cr$_{1.25}$O$_4$	8.325*	8.308
NiFe$_{0.5}$Cr$_{1.5}$O$_4$	8.322*	8.305
NiCr$_2$O$_4$	8.316	8.299

Note: * From Vegard's Law approximation

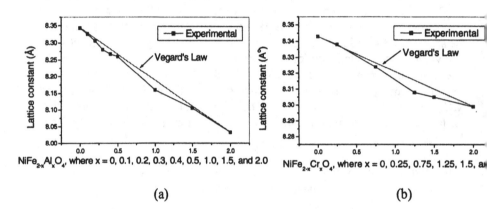

Figure 4. Variation of lattice parameters as a function of composition (a) NiFe$_{2-x}$Al$_x$O$_4$ system and (b) NiFe$_{2-x}$Cr$_x$O$_4$ system.

Density

In order to measure the electrical conductivity of pellets, the samples must be dense enough. Densities of more than 90 % of the theoretical value are considered satisfactory. Theoretical density (or XRD density) can be calculated (in g cm^{-3}) by the equation:[16]

$$d_{th} = \frac{zM}{0.6023V}$$

where M is molecular weight of the compound
z is the number of formula units per unit cell and
V is the volume of the crystalline unit cell as determined by X-ray diffraction (in Å^3)

Results of density measurements showed that all samples have densities above 90 % of the theoretical values. However, the density of spinel solid solution system $NiFe_{2-x}Al_xO_4$ and $NiFe_{2-x}Cr_xO_4$, sintered between 1400-1600 °C and 1300-1500 °C respectively, was found to decrease with increasing Fe content. This may be due to the loss of oxygen from samples or as a result of partial conversion of Fe^{3+} to Fe^{2+} at temperature in excess of 1300 °C.

Sensor Testing

The methane concentration dependence of impedance in $NiFe_2O_4$ and $NiCr_2O_4$ specimens was shown in figure 5 and 6. The impedance was found to increase as the methane concentration increased.

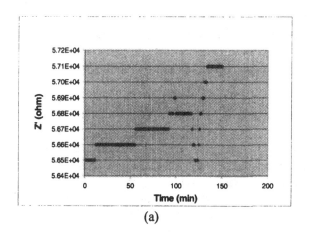

(a)

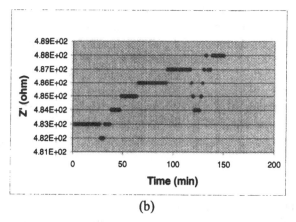

(b)

Figure 5. Response curves of impedance with CH_4 concentration ranging from 0 – 2.5 % over a period of 150 min in gaseous atmosphere (2.5 % CH_4 in air + Air) at the temperature 180 °C and frequency 1 kHz (a) $NiFe_2O_4$ sample (sintering temp. 1200 °C) and (b) $NiCr_2O_4$ sample (sintering temp. 1400 °C).

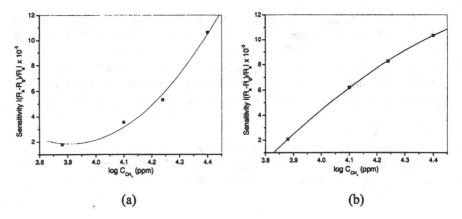

Figure 6. Sensitivity as a function of CH_4 concentration in gas mixture (2.5 % CH_4 in air + Air) at the temperature 180 °C and frequency 1 kHz (a) $NiFe_2O_4$ sample (sintering temp. 1200 °C) and (b) $NiCr_2O_4$ sample (sintering temp. 1400 °C).

CONCLUSION

In conclusion, the optimum calcining temperature to obtain nickel ferrite ($NiFe_2O_4$) and nickel aluminate ($NiAl_2O_4$) single phase was 1200 °C and 1400 °C respectively, whereas nickel chromite ($NiCr_2O_4$) single phase was completely formed at the same temperature as $NiAl_2O_4$, and they were all calcined for 3 hours. The complete formation of solid solutions of $NiFe_{2-x}Al_xO_4$ (where $x = 0.1, 0.2, 0.3, 0.4, 0.5, 1.0$, and 1.5) and $NiFe_{2-x}Cr_xO_4$ (where $x = 0.25, 0.75, 1.25$, and 1.50) occurred at 1400 °C after 3 hours. The XRD patterns of the solid solution system $NiFe_{2-x}Al_xO_4$ and $NiFe_{2-x}Cr_xO_4$ moved to higher values of 2θ when aluminium or chromium contents increased. $NiFe_{2-x}Al_xO_4$ and $NiFe_{2-x}Cr_xO_4$ formed a substitutional type of soild solutions, which was confirmed by the lattice parameter variation with increasing Al and Cr content of spinel solid solutions. This is probably due to the substitution of larger Fe^{3+} by smaller Al^{3+} and Cr^{3+} respectively. The pure spinels and their solid solutions were found to be sensitive to change in CH_4 concentration in the gas mixture (2.5 % CH_4 in air + Air) at 180 °C and frequency 1 and 10 kHz.

REFERENCES

[1] A.G. Salway, H.S. Eggleston, J.W.L. Goodwin, J.E. Berry, and T.P. Murrells, "UK Emissions of Air Pollutants 1970–1996," National Atmospheric Emissions Inventory, Department of the Environment, Transport and the Regions, 1999.

[2] P.T. Moseley and B.C. Tofield, "Appendix: Hazardous Concentrations of Various Gases in Air Mixtures"; in *Solid State Gas Sensors*, Adam Hilger, Bristol, 1987.

[3] L. de Angelis and R Riva, "Selectivity and Stability of a Tin Dioxide Sensor for Methane," *Sensors and Actuators B*, **28** 25–29 (1995).

[4] M. Fleischer and H. Meixner, "Sensitive, Selective and Stable CH_4 Detection Using Semiconducting Ga_2O_3 Thin Films," *Sensors and Actuators B*, **26-27** 81–84 (1995).

[5] G.K. Flingelli, M.M. Fleischer, and H. Meixner, "Selective Detection of Methane in Domestic Environments Using a Catalyst Sensor System Based on Ga_2O_3," *Sensors and Actuators B*, **48** 258–62 (1998).

[6] S.A. Akbar, C.C. Wang, L. Wang, and D.J. Collins, "Ceramic Oxides as Potential Hydrocarbon and NO_x Sensors"; pp. 331–42 in Ceramic Transactions, Vol. 65, *Role of Ceramics in Advanced Electrochemical Systems*. Edited by P.N. Kumta, G.S. Rohrer, and U. Balachandran. American Ceramic Society, Westerville, OH, 1996.

[7] A.D. Brailsford, M. Yussouff, and E.M. Logothetis, "Theory of Gas Sensors," *Sensors and Actuators B*, **13-14** 135–38 (1993).

[8] B.C. Tofield, "Current Technology-Semiconductor Sensors"; pp. 213–18 in *Solid State Gas Sensors*, Edited by P.T. Moseley and B.C. Tofield. Adam Hilger, Bristol, 1987.

[9] H.P. Klug and L.E. Alexander, "The Precision Determination of Lattice Constants"; pp. 594–597 in *X-Ray Diffraction Procedures for Polycrystalline and Amorphous Materials*, John Wiley & Sons, New York, 1974.

[10] J.B. Nelson and D.P. Riley, "An Experimental Investigation of Extrapolation Methods in the Derivation of Accurate Unit-Cell Dimensions of Crystals," *Proc. Phys. Soc. (London)*, **57** 160–77 (1945).

[11] G.M. Kale, "Thermodynamic Studies on Selected Ceramic Oxide Systems," Department of Metallurgy, Indian Institute of Science, Bangalore, 1990.

[12] L.B. Pankratz, J.M. Stuve, and N.A. Gocken, "Thermodynamic Properties of Elements and Oxides," Bulletin 672, United States Department of the Interior, Bureau of Mines, 1982.

[13] JCPDS-International Centre for Diffraction Data (ICDD), "Powder Diffraction File, Alphabetical Indexes, Inorganic Phases Sets 1–47," Pennsylvania, 1997.

[14] JCPDS-International Centre for Diffraction Data (ICDD), "Powder Diffraction File, Hanawalt Search Manual, Inorganic Phases Sets 1–47," Pennsylvania, 1997.

[15] R.D. Shannon and C.T. Prewitt, "Effective Ionic Radii in Oxides and Fluorides," *Acta Cryst.*, **B25** 925–46 (1969).

[16] S. Raman and J.L. Katz, in *Handbook of X-Rays*, Edited by E.F. Kaelble. McGraw-Hill, New York, 1967.

AMMONIA AND ALCOHOL GAS SENSORS USING TUNGSTEN OXIDE

Mana Sriyudthsak
Department of Electrical Engineering
Chulalongkorn University
Bangkok 10330, Thailand

Sophon Udomratananon
Department of Electrical Engineering
Chulalongkorn University
Bangkok 10330, Thailand

Sitthisuntorn Supothina
National Metal and Materials Technology Center,
National Science and Technology Development Agency
73/1 Rama 6 Rd., Rajdhevee, Bangkok 10400, Thailand

ABSTRACT

Tungsten oxide (WO_3) powders were prepared by precipitating from ammonium tungstate solution and then heat-treated at 800°C for 12 hours. Some powders were modified with 0.01 w% Au and 0.01 w% Cu. The gas sensors were fabricated from 3 types of powders : as-prepared, Au-modified and Cu-modified. Gas sensing performances to 0.01 - 100% alcohol and 0.01 - 10% ammonia were characterized by measuring the change of sensor's conductivity at operating temperature of 150 and 300°C. It was found that the non-modified WO_3 showed higher sensitivity to ammonia than to alcohol. Modification with 0.01% Au enhanced the sensitivity towards alcohol, while the powders modified with 0.01% Cu were sensitive to both alcohol and ammonia.

INTRODUCTION

There are needs of gas sensors which could be applied in severe environment. Among semiconductor materials, which are used to fabricate gas sensors, tungsten oxide is one of the most attractive materials for such applications.[1] The present paper describes the fabrication of gas sensor from tungsten oxide treated in various conditions.

MATERIALS AND METHOD

Tungsten oxide powders were prepared by precipitating from ammonium tungstate ($(NH_4)_{10}W_{12}O_{41} \cdot 5H_2O$, Wako, Japan) and nitric acid (HNO_3, Carlo) solution. The precipitates were separated, washed with de-ionized water for 3

To the extent authorized under the laws of the United States of America, all copyright interests in this publication are the property of The American Ceramic Society. Any duplication, reproduction, or republication of this publication or any part thereof, without the express written consent of The American Ceramic Society or fee paid to the Copyright Clearance Center, is prohibited.

times and dried at 120°C for 3 hours. The powders were then calcined at 800°C for 12 hours to form tungsten oxide (WO_3). Some powders were modified with gold and copper using gold colloid solution and copper sulfate solution, respectively, before subjected to calcination. The final content of the gold and copper dopants was 0.01 w%. Gas sensors were fabricated from 3 types of powders: as-prepared, Au-modified and Cu-modified. The powders were painted on titanium and platinum electrodes. Titanium and platinum with thickness of 50 and 100nm, respectively, were evaporated onto glass substrate to form electrodes. Gas sensing performances were characterized by measuring the conductivity change of the WO_3 layer when exposed to the sample gases. A measuring circuit was designed to eliminate the effect of the series resistance.[2] The sensors were set up in a flow through type measuring chamber. A carrier gas was a mixture of nitrogen and oxygen gases with N_2: O_2 volume ratio equal to 4:1 to obtain good response.[3] A 0.01 – 100% methyl alcohol and 0.01 - 10% ammonia were used as sample gasses. The samples were injected into the system in the form of solution (5 µL). A heater was installed at the injection port for evaporating the sample solution into gas phase during the measurement. The temperature of the injection port was set at 100°C, which was enough to evaporate the samples. The measurement was carried out at operating temperature of 150 and 300°C.

RESULTS AND DISCUSSIONS

From the experimental results, it was found that the responses at 300°C are better than that at 150°C.

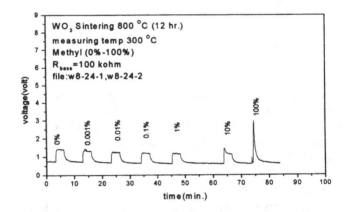

Figure 1. Response curve of the as-prepared WO_3 to methyl alcohol samples

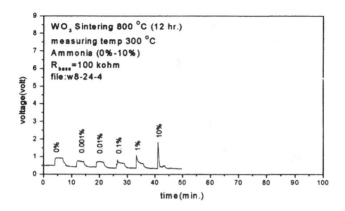

Figure 2. Response curve of the as-prepared WO_3 to ammonia samples

Figures 1 and 2 are the response curves of the as-prepared WO_3 sensor, measured at 300°C, when exposed to methyl alcohol and ammonia, respectively, at various concentrations. The responses of the Au-modified and Cu-modified WO_3 sensors are shown in figures 3 - 5. The peak and background voltages of the response curve were used to calculate a sensitivity of the sensors.

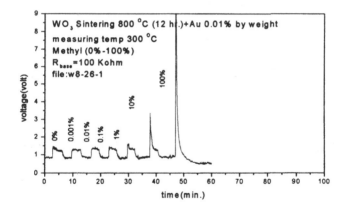

Figure 3. Response curve of the 0.01%Au modified WO_3 to methyl alcohol samples

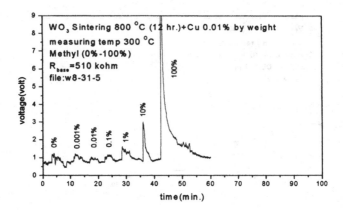

Figure 4. Response curve of the 0.01%Cu modified WO_3 to methyl alcohol samples

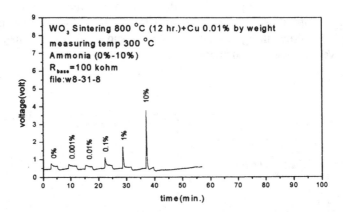

Figure 5. Response curve of the 0.01%Cu modified WO_3 to ammonia samples

The calibration curves of the sensors to methyl alcohol and ammonia are shown in figures 6 and 7. It is evident that the sensitivity of the WO_3 sensors towards methyl alcohol is enhanced by increasing operating temperature or by modifying the material with Au or Cu. Similarly, the sensitivity towards ammonia is improved when measured at 300°C or by modifying with Cu. However, modification with Au reduces the sensitivity.

With the above results, it could be summarized that the sensitivity and selectivity of tungsten oxide sensors could be enhanced by selecting appropriate operating temperature and/or modifying the raw material. By controlling these variables, it is possible to detect and classify volatile organic compounds such as alcohol and ammonia.

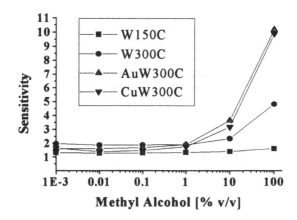

Figure 6. Calibration curves of the sensors to methyl alcohol
W150: WO_3 sensor at 150°C,
W300: WO_3 sensor at 300°C,
AuW300: Au modified WO_3 sensor at 300°C,
CuW300: Au modified WO_3 sensor at 300°C

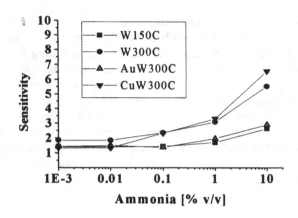

Figure 7. Calibration curves of the sensors to ammonia
W150: WO_3 sensor at 150°C,
W300: WO_3 sensor at 300°C,
AuW300: Au modified WO_3 sensor at 300°C,
CuW300: Au modified WO_3 sensor at 300°C

ACKNOWLEDGEMENT
This research is supported by the National Metal and Materials Technology Center, National Science and Technology Development Agency of Thailand.

REFERENCES
[1] Tomchenko, A., Emelianov, I., Khatko, V., "Tungsten trioxide-based thick film NO sensor: design and investigation" Sensors and Actuators B, 57 (1999) 166-170

[2] Teeramongkonrasmee A., Sriyudthsak M., "Problems in Gas Sensor Measuring Circuit and Proposal of New Circuit" Sensors and Materials, 11[3] (1999) 149-162

[3] Sriyudthsak M., Promsong L., Panyakeow S., "Effect of Carrier Gas on Response of Oxide Semiconductor Gas sensor", Sensors and Actuators B, 13-4 (1993) 139-142

LOW TEMPERATURE GAS SENSING USING LASER ACTIVATION

Mana Sriyudthsak
Department of Electrical Engineering
Chulalongkorn University
Bangkok 10330 Thailand

Voratat Rungsaiwatana
Department of Electrical Engineering
Chulalongkorn University
Bangkok 10330 Thailand

ABSTRACT

Tin oxide (SnO_2) powder was prepared by precipitating from stannic chloride ($SnCl_4$) and ammonium hydroxide solution and then calcined at 600 and 800°C for 12 hours. The powder was used to fabricate gas sensors. Gas sensing performances were characterized by measuring the change of sensor's conductivity at operating temperature of 50, 80 and 120°C under conventional condition and laser activation. Alcohol and ammonia were used as sample gases. It was found that the higher sensitivity and the better signal to noise ratio of the responses were obtained under the laser activation, especially when the 600 °C calcined tin oxide was used at the operating temperature of 50 °C.

INTRODUCTION

One of the main problems of semiconductor gas sensors is the requirement of high operating temperature. This dues to the low conductivity and low carrier concentration of the semiconductor materials at low temperature. This limits the application of semiconductor gas sensors in an explosive environment. In this paper, we propose a technique for reducing the operating temperature of the semiconductor gas sensor by using laser stimulation.

MATERIALS AND METHODS

The semiconductor gas sensors were prepared from tin oxide (SnO_2). The tin oxide powder was prepared using precipitation method from stannic chloride ($SnCl_4$) and ammonium hydroxide solution. The powder was then calcined at 600 and 800°C for 12 hours. The powder was painted on to electrodes to form gas sensors. The electrodes were fabricated by evaporating titanium and platinum with thickness of 50 and 100nm on to glass substrates, respectively. The sensor was set in to a measuring chamber with a transparent quartz window, which allows the laser beam passes and arrives the sensor surface. Beneath the sensor, a heater was installed for controlling the operating temperature of the sensor. The temperature was varied from 50 to 120 °C. A carrier gas was a mixture of nitrogen and oxygen gases with N_2: O_2 volume ratio equal to 4:1 to obtain good response.[1] The flow rate of the nitrogen and oxygen were 40 and 10 ml/min, respectively. The conductivity of the sensors was measured for investigating the sensor performances using measuring circuit. The circuit was designed to eliminate the effect of the series resistance.[2] Figure 1 shows the schematic diagram of the measuring system. Nitrogen laser with wavelength of 337nm was used in stimulating the gas sensors during the measurements.

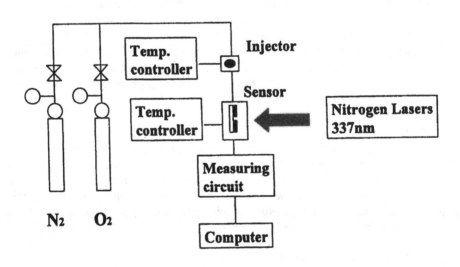

Figure 1. Schematic diagram of the measuring system

RESULTS AND DISCUSSION

Figure 2 shows a typical response curves of the 600 °C, 12 hours calcined SnO_2 sensors when they were exposed to alcohol samples at operating temperature of 50°C under normal and laser stimulation. It is obvious that when there was no laser stimulation the signal was noisy. However under the laser stimulation, the signal with a better signal to noise(S/N) ratio could be obtained. The amplitudes of the responses under laser stimulation were also higher than the normal one. The response curves of the same sensor at operating temperature of 120°C are shown in figure 3. The noisy signals of the normal condition were now disappeared. It could be observed that the recovery time of the sensors under laser stimulation was faster than the normal one.

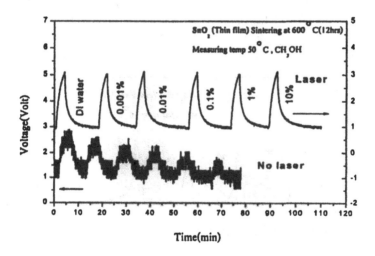

Figure 2. Response curves of the 600 °C–calcined SnO_2 sensors to alcohol at operating temperature of 50°C under normal and lasers stimulating condition

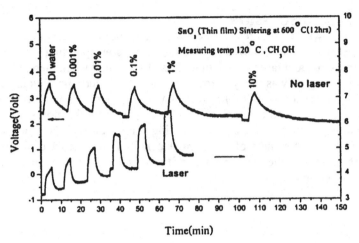

Figure 3. Response curves of the 600 °C–calcined SnO_2 sensors to alcohol at operating temperature of 120°C under normal and lasers stimulating condition.

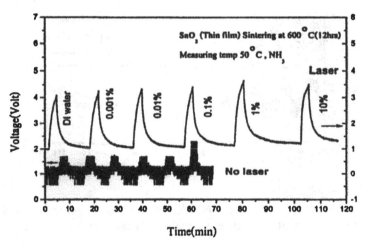

Figure 4. Typical response curves of the 600 °C–calcined SnO_2 sensors to ammonia at operating temperature of 50°C under normal and lasers stimulating condition

The same trend was also observed when the sensors were exposed to ammonia as shown in figures 4 and 5.

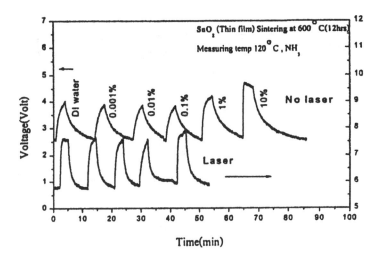

Figure 5. Typical response curves of the 600 °C–calcined SnO_2 sensors to ammonia at operating temperature of 120°C under normal and lasers stimulating condition

Figure 6 shows the calibration curves of the sensors to alcohol at operating temperature of 50°C. It was found that there was an increasing of the sensor sensitivity for the 600°C-calcined sensors, but there was almost no significant for the 800 °C-calcined sensors. This may be due to the different microstructure of the sensors when they were calcined at different temperature. However from figure 6, the calibration curves were not so good since the slope of the curves were not clear and high enough to estimate the sample concentration. This problem could be solved by modifying the sensors with an appropriated metal. Figure 7 shows the calibration curves of the sensors to alcohol at operating temperature of 120°C. All sensors, which were calcined at 600 and 800 °C under normal and laser stimulation, had nearly the same characteristics. The effect of the laser stimulation at this high temperature could not be observed.

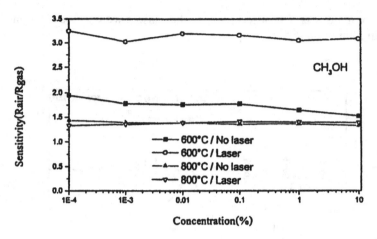

Figure 6. Calibration curves of the sensors to alcohol operating at 50°C.

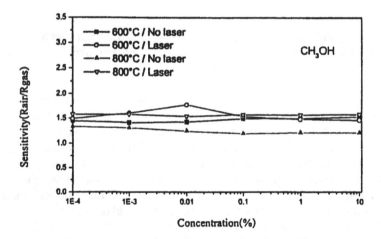

Figure 7. Calibration curves of the sensors to alcohol operating at 120°C.

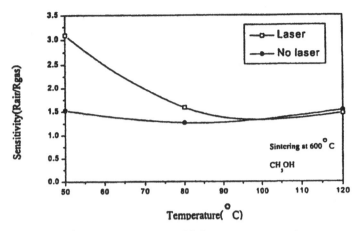

Figure 8. Relation between the sensitivity and the operating temperature

The relation between the sensitivity and the operating temperature is shown in figure 8. It could be observed that the sensitivity of the sensor increased when it was operated at low temperature only under laser stimulation. For the higher temperature operation, the sensitivity was not significantly increased. This could be explained that at lower temperature and not lasers stimulating condition, the sensors might have not enough carriers for the signal transduction. However, when the sensors was under lasers stimulation, there was an increasing of carrier concentration in the sensors. This is because of that the nitrogen lasers with wavelength of 337nm(~3.7eV) has enough energy for stimulating the electron from the valence band to the conduction band of the SnO_2 sensor (band gap ~3.3eV). However, at higher temperature, they may have enough carriers for the signal generation, then the effect of the laser stimulation could not be observed. These results indicate a feasibility of using gas sensor at a low temperature.

REFERENCES

[1] Sriyudthsak M., Promsong L., Panyakeow S., "Effect of Carrier Gas on Response of Oxide Semiconductor Gas sensor", Sensors and Actuators B, 13-4 (1993) 139-142

[2] Teeramongkonrasmee A., Sriyudthsak M., "Problems in Gas Sensor Measuring Circuit and Proposal of New Circuit" Sensors and Materials, 11[3] (1999) 149-162

SYNTHESIS OF GALLIUM OXIDE HYDROXIDE CRYSTALS IN AQUEOUS SOLUTIONS WITH OR WITHOUT UREA AND THEIR CALCINATION BEHAVIOR

A. Cüneyt Taş
Merck Biomaterial GmbH
Frankfurterstr. 250, F129/218
Darmstadt 64293, Germany

Peter J. Majewski and Fritz Aldinger
Max-Planck-Institut für Metallforschung
Heisenbergstr. 5
Stuttgart 70569, Germany

ABSTRACT

Gallium oxide hydroxide (GaOOH·xH$_2$O) single crystals were synthesized in aqueous solutions by using two different precipitation techniques; homogeneous decomposition of urea and forced hydrolysis in pure water. Precipitation of crystals started at exactly the same pH value (i.e., 2.05 at 85°C) in both cases. The morphology of crystals turned out to be quite different (zeppelin-like with urea, rod-like without urea) in each of the above methods. Calcination of these gallium oxide hydroxide crystals in air at temperatures ≥500°C transformed them into Ga$_2$O$_3$. Characterization of the samples were performed by x-ray diffraction, scanning electron microscopy, thermal analyses, infrared spectroscopy, and carbon and nitrogen analyses.

INTRODUCTION

Gallium (named after the region Gallia of France[1]) is an element, which is more abundant in the earth's crust as compared to some of the technologically important elements, such as B, Pb, Bi, Nb, Mo, W, Hg, or Sn[2]. Gallium oxide, like other oxides of Group III metals, is widely used for the preparation of phosphors and catalysts. Ga$_2$O$_3$ is normally an insulator, with a forbidden energy gap of ~4.9 eV[3]. However, calcination of Ga$_2$O$_3$ in reducing atmospheres turns it into an n-type semiconductor, due to the creation of oxygen vacancies[4]. The high-temperature structure of Ga$_2$O$_3$ (β form) is monoclinic (space group C2/m), and this is only one of five well known forms[5] of gallium oxide: α-, χ-, δ-, and ε-Ga$_2$O$_3$, and all of these polymorphs are converted to β-Ga$_2$O$_3$ at T>870°C[6]. Manufacture of *n*-type semiconducting Ga$_2$O$_3$ thin films on electrically insulating substrates have been studied by Meixner, *et al.*[7], which are considered and developed for the detection of reducing gases. It was also reported that different gas sensitivities can be chosen by the appropriate setting of the operating temperature, and at temperatures in excess of 900°C, the sensors may even be operated as O$_2$ sensors[8]. On the other hand, Haneda, *et al.* reported[9] that the Ga$_2$O$_3$-Al$_2$O$_3$

ceramic catalysts prepared by using the sol-gel method showed superior activity (as compared to those of either pure Al_2O_3 or Ga_2O_3) for the selective reduction of NO with propene in the presence of H_2O and SO_2.

Gallium oxide has recently been used in the synthesis of solid electrolytes of superior (as compared to stabilized zirconia's) ionic conductivity, i.e., $La_{0.8}Sr_{0.2}Ga_{0.8}Mg_{0.2}O_{2.8}$ (LSGM)[10], also beyond its significance for the semiconductor, optoelectronic, and catalysis technologies. We have encountered the fortuitous formation of GaO(OH) in aqueous solutions throughout the course of our recent studies, which were mainly focused on the wet-chemical synthesis of doped $LaGaO_3$ fuel cell ceramics[11, 12].

Laubengayer, *et al.*[13] synthesized α-GaO(OH) from a gel obtained *via* hydrolysis of gallium nitrate (or gallium chloride) at temperatures between 110° and 300°C. They found that $Ga(OH)_3$ transformed slowly into α-GaO(OH) in the aqueous solution. Precipitation of GaO(OH) from $GaCl_3$ solutions, upon the addition of various alkalis (such as NaOH and KOH), has also been studied by Sato, *et al.*[14], and they reported that the freshly precipitated (at pH values varying between 6 and 10) precursors were all X-ray amorphous and converted to crystalline α-GaO(OH), with an orthorhombic crystal structure similar to that of diaspore (α-AlO(OH))[15], only after about 1 day of aging in their mother liquors. α-GaO(OH) is a structural analogue of other oxidic hydrates, such as α-MnO(OH)[16] and Goethite: α-FeO(OH)[17]. Avivi, *et al.*[18] produced tubular particles of α-GaO(OH), after inserting an ultrasonic finger into 0.114 M aqueous solutions of $GaCl_3$ for 6 h. They claimed that the sonochemical reaction provided by the ultrasonic finger caused the hydrolyses of Ga-chloride solutions.

In this study, we used the precipitation methods of (a) the homogeneous decomposition of urea[19-21], and (b) forced hydrolysis (without urea, at 90°C) of dilute gallium nitrate solutions to form the GaO(OH) crystals. We believe that this study will help the upcoming researchers who would work on the wet-chemical synthesis of Ga-containing ceramics, which may have numerous applications.

EXPERIMENTAL

$Ga(NO_3)_3 \cdot 4.06H_2O$ (99.999%, Sigma-Aldrich Chemie GmbH, Steinheim, Germany) was used as the gallium source. A 0.37 M stock solution was first prepared by dissolving an appropriate amount of gallium nitrate in de-ionized water. Urea used was also reagent-grade (>99.5%, Merck GmbH, Darmstadt, Germany). Synthesis experiments were performed by two different precipitation techniques. For the homogeneous precipitation experiments, a 6.7 mL aliquot of the Ga-stock solution was mixed with 90 mL of de-ionized water, and 1.756 g of urea was then added. The resultant solution was heated on an hot-plate to 90°C in about 45 minutes. For the forced hydrolysis experiments, again a 6.7 mL aliquot of the Ga-stock solution was mixed with 90 mL of de-ionized water, followed by heating it on an hot-plate to 90°C in about 45 minutes. Precipitates (in each of the above-described procedures), after 95 minutes of aging at the constant temperature of 90°C, were finally separated from their mother liquors by centrifugal filtration, followed by washing 4X with 2-propanol (>99%, Merck). Washed precipitates were dried in an oven at 90°C, overnight. The dried powders were first finely ground by using an agate mortar/pestle, and then calcined for 6 h (with a heating and cooling rate of 5°C/min), as loose powders, in alumina boats in air at various temperatures in the range of 250° to 1200°C.

Phase constitution of the powders was investigated by a powder X-ray diffractometer (D-5000, Siemens GmbH, Karlsruhe, Germany) using CuK_α radiation (40kV, 30 mA, step size: 0.016°, count time: 1 s). Pyrolysis of the accurately-weighed 224.8 mg portions of powders was monitored by simultaneous differential thermal and thermogravimetric analysis (STA501, Bähr GmbH, Bremen, Germany) in air with a scan rate of 5°C/min. Powder samples (1 wt% in KBr) were also characterized by FTIR (IFS 66, Bruker GmbH, Karlsruhe, Germany). C and N contents of samples were determined by the combustion-IR absorption method (CS-800, Eltra GmbH, Neuss, Germany). Powder samples were also investigated by field-emission scanning electron microscopy (FESEM: DSM 982 Gemini, Zeiss GmbH, Oberkochen, Germany).

RESULTS AND DISCUSSION

Both techniques used in this study produced well defined, monodisperse crystals of $GaO(OH) \cdot xH_2O$. Variation of pH, with time, during the precipitation runs (for both routes) has been continuously monitored and a

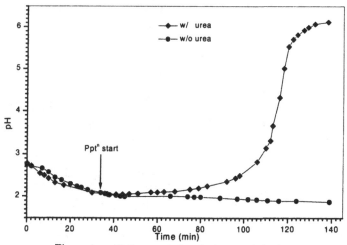

Figure 1 pH-time variation during precipitation runs

typical plot is given in Figure 1. Thermal analysis of the precipitates of both

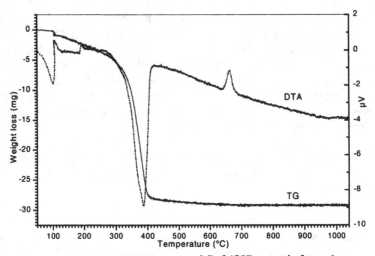

Figure 2   TG/DTA traces of GaO(OH) crystals formed

routes produced the same TG/DTA data and they are given in Figure 2. Precipitation started (seen by the onset of turbidity in the solutions) when the temperature reached 85°-86°C, regardless of the presence of urea in the solutions. The initial pH value of the 0.026 M Ga^{3+} solutions (with or without urea) was in the vicinity of 2.80 at 23°C, and precipitation started in both kinds of solutions when pH dropped down exactly to 2.05.

On the other hand, the steep rise in pH (when t > 80 min) observed in urea-containing solutions is characteristic for the decomposition of urea. The detailed sequence of reactions describing the decomposition of urea in aqueous media has previously been documented[22].

GaO(OH) crystals displayed (Figure 2) a total weight loss of about 13% (up to 1040°C). It may be stated that the chemical formula of the freshly precipitated crystals were not simply GaO(OH), but it is rather GaO(OH)·$0.2H_2O$, and these powders readily converted into Ga_2O_3 upon heating to ≥400°C. Dried (90°C) precipitates, as well as those calcined at 250°C (Figure 3) had the orthorhombic structure of GaO(OH) with the lattice parameters, where a = 4.5606, b = 9.7975, c = 2.9731 Å, and V = 132.85 Å^3. Powders heated at 500°C, for 6 h, showed the typical XRD trace of the rhombohedral polymorph of Ga_2O_3 (ICDD PDF 06-0503). The only phase identified in the XRD spectra of 750°, 1000°, and 1300°C was the monoclinic polymorph of Ga_2O_3 (PDF 41-1103 or 76-0573). The broad endothermic peak appeared between 290° and 400°C was due to the elimination of water from

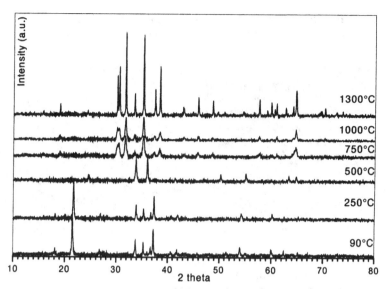

Figure 3   XRD traces of GaO(OH) crystals as a function of temperature

constitutional OH groups (Figure 2). The exothermic peak detected in the DTA trace of the powders, at around 670°C, corresponded to the polymorphic phase transformation in Ga_2O_3, from rhombohedral to monoclinic. The FTIR spectra (Figure 4) of the 90°C and 250°C-samples also showed (in agreement with the XRD data) the bands at 2036 and 1942, together with the bands at 1026 and 952 cm^{-1}, and these were assigned to be the constitutional Ga-OH bending bands, and their overtones, respectively, as reported by Sato, et al.[14].

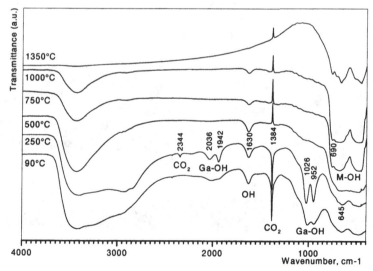

Figure 4   FTIR spectra of GaO(OH) crystals as a function of temperature

The main difference between the two synthesis techniques of this study actually came in terms of the morphology of the precipitates produced.

Figure 5   Recovered GaO(OH) precipitates; (*Left*) w/o urea, (*Right*) w/urea

GaO(OH) precipitates (Figure 5) formed in pure water (i.e., w/o urea) were recovered from their mother liquors at the end of 95 minutes of aging at 90±1°C and at the final pH value of 1.87. These single crystals possessed a unique rod-like morphology as shown above, with an average rod length of about 3 µm. However, the precipitates formed just at the precipitation-start point (i.e., pH=2.05, T=86°C) in urea-containing solutions consisted of little (200 to 500 nm-long) crystalline "zeppelins." Precipitates recovered from the same solution (w/urea) after 95 minutes of aging (at 90±1°C, final pH=6.10) in the mother liquor still had the unique zeppelin morphology, but their lengths increased to about 1 to 2 µm. Bigger zeppelins were also crystalline and had the same orthorhombic unit cell as given above. This unique morphology (zeppelins or elongated acicular or epitaxial-twinned crystals) observed here for GaO(OH), which were only encountered in urea-containing solutions, has also previously been encountered by Parida, et al.[17] and Goni-Elizalde, et al.[23] during the synthesis of α-FeOOH in the presence of urea.

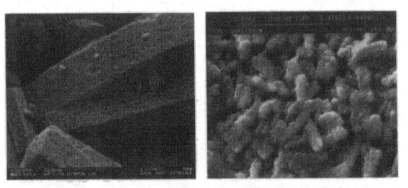

Figure 6   Monoclinic Ga_2O_3 (calcined at 1000°C); (*Left*) w/o urea, (*Right*) w/urea

Calcination of the powders at 1000°C in air, for 6 h, also produced a difference in their morphology as shown in Figure 6. Rods (obtained by forced hydrolysis) of monoclinic Ga_2O_3 preserved their initial shape, but developed nano-scale porosity on their surfaces. On the other hand, the initial zeppelin-like morphology totally disappeared, upon heating the powders obtained via homogeneous decomposition of urea at 1000°C, although they had the characteristic XRD pattern for monoclinic Ga_2O_3.

Avivi, et al.[18] reported the strong necessity of "ultrasonic hydrolysis for the sonochemical formation" of scroll-like cylindrical or tubular GaO(OH) particles. It should be remembered that the insertion of an ultrasonic finger into a solution, and operating it (@100 W/cm^2) in that solution for a long time like 6 hours would certainly cause a lot of heating of that solution. What these authors actually observed at the end of 6 hours could simply be the forced hydrolysis of a gallium solution at an elevated temperature.

Figure 7 Monoclinic Ga_2O_3 (*L*) w/o urea, 1000°C, (*R*) commercial Ga_2O_3 powders

As seen from the comparative micrographs of Figure 7, GaO(OH) powders obtained in this study, for instance, by forced hydrolysis may easily be converted to monoclinic Ga_2O_3 after heating at 1000°C, and they possess more uniform morphology as compared to those of Sigma-Aldrich Ga_2O_3 powders (Lot. No. 33281-110).

CONCLUSIONS

Upon heating a dilute (0.026 M) aqueous solution of Ga-nitrate at 90°C for 95 minutes (i.e., forced hydrolysis), rod-like, monodispersed, 1-3 µm-long orthorhombic GaOOH·0.2H$_2$O single crystals can be formed (starting at the pH value of 2.05 at 85°C), whereas upon heating a dilute (0.026 M) solution of Ga-nitrate, which also contains 0.304 M urea (i.e., homogeneous decomposition of urea), to 90°C, zeppelin-shaped, monodispersed, 200-500 nm-long, orthorhombic GaOOH·0.1H$_2$O single crystals can be formed, again at the exact pH value of 2.05. GaO(OH) crystals transform first into rhombohedral and then monoclinic polymorphs of Ga_2O_3, upon calcination in an air atmosphere for 6 h at temperatures $\geq 500°C$. Carbon and nitrogen analyses, as well as the FTIR results, showed that the formed crystals (following drying at 90°C) did not contain any structural CO_3^{2-} and NO_3^- ions.

ACKNOWLEDGMENTS

One of the authors, Dr. A. Cüneyt Taş, gratefully acknowledges the Max-Planck-Institut für Metallforschung, Stuttgart for the award of Visiting Professorship, extending from February 1999 to February 2001. The authors also express their gratitude to M. Thomas (XRD), H. Labitzke (FESEM), H. Kummer (TG/DTA), W. König (FT-IR), F. Predel (SEM), and S. Hammoud (C & N analyses) for their generous help on sample characterization.

REFERENCES

[1] R. C. Weast, D. R. Lide, M. J. Astle, and W. H. Beyer (Eds). *CRC Handbook of Chemistry and Physics*; p. B-17. CRC Press, Inc., 70th Edition, Boca Raton, Florida, 1990.

[2] C. Klein and C. S. Hurlbut, *Manual of Mineralogy*; pp.222-3. John Wiley Sons, Inc. New York, 1993.

[3] L. Binet, D. Gourier, and C. Minot, "Relation between Electron Band Structure and Magnetic Bistability of Conduction Electrons in β-Ga_2O_3," *J. Sol. State Chem.*, **113**, 420-33 (1994).

[4] L. Binet and D. Gourier, "Origin of the Blue Luminescence of β-Ga_2O_3," *J. Phys. Chem. Solids*, **59**, 1241-49 (1998).

[5] R. Roy, V. G. Hill, and E. F. Osborn, "Polymorphism of Ga_2O_3 and the System of Ga_2O_3-H_2O," *J. Am. Chem. Soc.*, **74**, 719-22 (1952).

[6] M. Fleischer, L. Höllbauer, E. Born, and H. Meixner, "Evidence for a Phase Transition of β-Gallium Oxide at Very Low Oxygen Pressures," *J. Am. Ceram. Soc.*, **80**, 2121-25 (1997).

[7] M. Fleischer and H. Meixner, "Sensing Reducing Gases at High Temperatures using Long-term Stable Ga_2O_3 Thin Films," *Sensors and Actuators*, **B6**, 277-81 (1992).

[8] J. Frank, M. Fleischer, and H. Meixner, "Gas-sensitive Electrical Properties of Pure and Doped Semiconducting Ga_2O_3 Thick Films," *Sensors and Actuators*, **B48**, 318-21 (1998).

[9] M. Haneda, Y. Kintaichi, H. Shimada, and H. Hamada, "Selective Reduction of NO with Propene over Ga_2O_3-Al_2O_3: Effect of Sol-Gel Method on the Catalytic Performance," *J. Catal.*, **192**, 137-48 (2000).

[10] T. Ishihara, H. Matsuda, and Y. Takita, "Doped $LaGaO_3$ Perovskite Type Oxide as a New Oxide Ionic Conductor," *J. Am. Chem. Soc.*, **116**, 3801-3 (1994).

[11] A. C. Taş, P.J. Majewski, and F. Aldinger, "Chemical Preparation of Pure and Strontium- and/or Magnesium-doped Lanthanum Gallate Powders,," *J. Am. Ceram. Soc.*, **83**, 2954-60 (2000).

[12] A. C. Taş, P. J. Majewski, and F. Aldinger, "Preparation of Sr- and Zn-Doped LaGaO$_3$ Powders by Precipitation in the Presence of Urea and/or Enzyme Urease," *J. Am. Ceram. Soc.*, In Print (2002).

[13] A. W. Laubengayer and H. R. Engle, "The Sesquioxide and Hydroxide of Gallium," *J. Am. Chem. Soc.*, **61**, 1210-4 (1939).

[14] T. Sato and T. Nakamura, "Studies of the Crystallisation of Gallium Hydroxide Precipitated from Hydrochloric Acid Solutions by Various Alkalis," *J. Chem. Tech. Biotechnol.*, **32**, 469-75 (1982).

[15] A. Klug and L. Farkas, "Structural Investigations of Polycrystalline Diaspore Samples by X-Ray Powder Diffraction," *Phys. Chem. Minerals*, **7**, 138-40 (1981).

[16] L. S. D. Glasser and L. Ingram, "Refinement of the Crystal Structure of Groutite, α-MnOOH," *Acta Cryst.*, **B24**, 1233-6 (1968).

[17] K. Parida and J. Das, "Studies on Ferric Oxide Hydroxides. 2. Structural Properties of Goethite (α-FeOOH) Prepared by Homogeneous Precipitation from Fe(NO$_3$)$_3$ Solution in the Presence of Sulfate Ions," *J. Coll. Int. Sci.*, **178**, 586-93 (1996).

[18] S. Avivi, Y. Mastai, G. Hodes, and A. Gedanken, "Sonochemical Hydrolysis of Ga^{3+} Ions: Synthesis of Scroll-like Cylindrical Nanoparticles of Gallium Oxide Hydroxide," *J. Am. Chem. Soc.*, **121**, 4196-9 (1999).

[19] E. Matijevic, "Monodispersed Colloids-Art and Science," *Langmuir*, **2**, 12-20 (1986).

[20] A. C. Taş, "Preparation of Lead Zirconate Titanate (Pb(Zr$_{0.52}$Ti$_{0.48}$)O$_3$) by Homogeneous Precipitation and Calcination," *J. Am. Ceram. Soc.*, **82**, 1582-4 (1999).

[21] A. Kato and Y. Morimitsu, "Formation of Iron(III) Hydroxide Particles from Iron(III) Salt Solutions by Homogeneous Precipitation Method," *Nippon Kagaku Kaishi*, **6**, 800-7 (1984).

[22] D. J. Sordelet, M. Akınç, M. L. Panchula, Y. Han, and M. H. Han, "Synthesis of Yttrium-Aluminum-Garnet Precursor Powders by Homogeneous Precipitation," *J. Eur. Ceram. Soc.*, **14**, 123-30 (1994).

[23] S. Goni-Elizalde and M. E. Garcia-Clavel, "Thermal Behaviour in Air of Iron Oxyhydroxides Obtained from the Method of Homogeneous Precipitation. Part II. Akaganeite Sample," *Thermochim. Acta*, **129**, 325-334 (1988).

KEYWORD AND AUTHOR INDEX

2–3 spinels, 79

Akbar, S., 37, 67
Alcohol gas sensor, 91
Aldinger, F., 105
Ammonia gas sensor, 91
Antimony sensor, 57
Automotive applications, 1

Blau, P., 37
Brosha, E.L., 1
Brydson, R.M.D., 79

Calcination, 105
Calcium zirconate, 47
Cao, W., 67
Conductivity, 57
Copper dopant, 91

Density, 57
Diffusion equation, 37

Ejakov, S.G., 11
Electrochemical sensor, 1
Ethanol sensor, 67

Feng, B., 37
Fergus, J.W., 47

Gallium oxide hydroxide, 105
Garzon, F.H., 1
Gas sensor, durability of, 37
Gold dopant, 91

Hostile exhaust environments, 19
Hydrocarbons, 1
Hydrogen sensor, 47

Iron oxide, 67

K-β-Al_2O_3, 57
Kale, G.M., 57, 79
Kubinski, D.J., 11
Kurchania, R., 57

Lanthanum chromate, 1
Low-temperature gas sensing, 97

Majewski, P.J., 105
Matsunaga, N., 27
Merhaba, A., 37
Methane sensor, 79
Mixed potential, 1, 11
Morphology, 105
Mukundan, R., 1

Nanopowders, 67
Nelson, C.S., 19
Newaz, G., 37
Nickel aluminate, 79
Nickel chromite, 79
Nickel ferrite, 79
Nietering, K.E., 11
Nitrogen laser, 97
NO_x sensor, 11

Packaging, 19
Parsons, M.H., 11

Planar exhaust sensor, 19
Poomiapiradee, S., 79

Reddy, C.V.G., 67
Riester, L., 37
Rungsaiwatana, V., 97

Sazai, G., 27
Semiconductor gas sensor, 37
Setiawan, A.H., 47
Solid electrolytes, 47, 57
Solid solutions, 79
Soltis, R.E., 11
Sriyudthsak, M., 91, 97
Supothina, S., 91
Synthesis, 105

Tan, O.K., 67
Taş, A.C., 105
Thick films, adhesion of, 37
Thin films, 37
Tin oxide, 97
Tungsten oxide, 91

Udomratananon, S., 91
Urea, 105

Visser, J.H., 11

Yamazoe, N., 27
Yttria-stabilized zirconia, 1

Zhu, W., 67
Zirconia, 11, 67